建筑立场系列丛书 No.67

木材——诗意与实用
Wood
Poetic and Practical Possibilities

福斯特建筑事务所等 | 编
刘文静 | 译

大连理工大学出版社

木材
——诗意与实用

- 004 **木材——诗意与实用** _ Susanne Kennedy
- 012 ICD/ITKE 研究馆2015-16 _ ICD + ITKE + University of Stuttgart
- 022 曼彻斯特的新麦姬癌症中心 _ Foster + Partners
- 036 圆形木亭 _ Christiansen Andersen
- 046 Caltron的新型社区 _ Mirko Franzoso Architetto
- 062 Hacine Cherifi体育馆 _ Tectoniques Architects
- 078 塞拉基乌斯博物馆扩建项目 Gösta馆 _ MX_SI
- 096 Knarvik社区教堂 _ Reiulf Ramstad Arkitekter
- 114 Créteil教堂的扩建项目 _ Architecture-Studio
- 126 苏黎世动物园大象馆 _ Markus Schietsch Architekten
- 142 Sognefjellshytta高山酒店的新入口 _ Jensen & Skodvin Architects

- 158 **扎哈·哈迪德：光辉人物的精神遗产** _ Silvio Carta + Marta González
- 170 萨勒诺海运码头 _ Zaha Hadid Architects

- 184 建筑师索引

Wood
Poetic and Practical Possibilities

004 *Wood – Poetic and Practical Possibilities_Susanne Kennedy*

012 ICD/ITKE Research Pavilion 2015-16_ICD+ITKE+University of Stuttgart

022 New Maggie's Cancer Center in Manchester_Foster+Partners

036 Around Pavilion_Christiansen Andersen

046 New Social House in Caltron_Mirko Franzoso Architetto

062 Hacine Cherifi Gymnasium_Tectoniques Architects

078 Serlachius Museum Extention, Gösta Pavilion_MX_SI

096 Community Church Knarvik_Reiulf Ramstad Arkitekter

114 Créteil Cathedral Expansion_Architecture-Studio

126 Elephant House Zoo Zürich_Markus Schietsch Architekten

142 Sognefjellshytta High Mountain Hotel's New Entrance_Jensen & Skodvin Architects

158 *Zaha Hadid: The Legacy of a Star_Silvio Carta+Marta González*

170 Salerno Maritime Terminal_Zaha Hadid Architects

184 Index

木材——诗意与实用
Poetic and Practical Possibilities

作为建筑材料的一种，木材在全球掀起了一股复兴的潮流。这股潮流的形成很大部分取决于在环保意识盛行的时代，碳化木所具有的独一无二的地位。近期的建筑项目从各个方面很好地佐证了木材所具备的价值，不仅富有诗意而且极具可行性，因此木材被很多人比作是21世纪的"混凝土"。

十座以居民和社区为设计重点的建筑物将处在人们的检验之中。这些建筑采用的都是具有地方性特点的标志性木材。颇具特色的木材构成了项目的核心，呈现在富有特色的私人住宅和公共建筑中。同时这些木材的存在，也对于每个设计项目与其周围景观的结合程度以及对其周围景观的参照程度都起到了影响。

这一引人注目的建筑系列包含了朝圣、休闲空间以及市民公共区域，每座建筑因为其单一的材料和设计方式在一定程度上或多或少地焕发起人们自信、优雅以及镇定的感觉。

现在让我们将注意力集中在光线、重复性和利用性的相互结合、玻璃的巧妙利用、体量的有效控制、创新型的屋面设计以及迷宫理念的采用方面。这些都是用来促进建筑物与自然环境的结合，并探索其中奥秘的方法。

Wood is undergoing a global revival as a construction material, partly due to its status as an unrivaled carbon sequesterer in an age of environmental awareness. Recent architectural projects exemplify its diverse poetic and practical attributes, prompting some to describe timber as the concrete of the 21st century.

Ten civic and community-focused structures, which use regional, native timbers as their signature material will be examined, with particular focus on the characteristics wood bestows on these projects, collectively and individually, and with what degree each design references or integrates with the surrounding landscape.

While this striking architectural collection includes places of worship, recreational, civil edifices, each to a greater or lesser extent evokes a sense of stillness, elegance and calm, for their stripped-back material pallets and design approach.

Attention to light, interplay between repetition and diversity, a deft use of glass, manipulation of volumes, innovative roofing solutions, and employment of the maze concept are some of the methods that have been utilized to facilitate a connection between and discovery of built and natural environments.

ICD/ITKE研究馆2015-16_ICD/ITKE Research Pavilion 2015-16/ICD+ITKE+Stuttgart University
曼彻斯特的新麦姬癌症中心_New Maggie's Cancer Center in Manchester/Foster+Partners
圆形木亭_Around Pavilion/Christiansen Andersen
Caltron的新社区_New Social House in Caltron/Mirko Franzoso Architetto
Hacine Cherifi体育馆_Hacine Cherifi Gymnasium/Tectoniques Architects
塞拉基乌斯博物馆扩建项目——Gösta馆_Serlachius Museum Extention, Gösta Pavilion/MX_SI
Knarvik社区教堂_Community Church Knarvik/Reiulf Ramstad Arkitekter
Créteil教堂的扩建项目_Créteil Cathedral Expansion /Architecture-Studio
苏黎世动物园大象馆_Elephant House Zoo Zürich/Markus Schietsch Architekten
Sognefjellshytta高山酒店的新入口_Sognefjellshytta High Mountain Hotel's New Entrance/Jensen&Skodvin Architects

木材——诗意与实用_Wood – Poetic and Practical Possibilities/Susanne Kennedy

过去十年来，木材在建筑行业又一次兴起，主要原因是工程木产品进行了改良，工程木其实早在20世纪初就出现过。因此，在相对较短的时间内，一系列新型木材产品实现了从小众产品到标准材料的过渡，为众多建筑者和设计者所熟知。这一系列的工程木产品中，"交叉层压木材，也就是人们所知悉的胶合木[1]，被认为是街区建筑材料中的新面孔"。[2]

新的电脑技术的应用意味着木材产品可以经过加工制成越来越多的复杂形式、更长的长度和更宽的宽度，且加工时间会越来越快，同时价格也会越来越低廉。[3] 木材的改良以及木材本身所具备的轻巧性、强度和重量的比率，使它同样适用于预制加工和屋面设计，也更加自信地、创新地应用在大规模的结构之中。[4] 本期封面所展示的斯图加特大学ICD/ITKE研究馆，揭示了众多尖端技术所蕴含的巨大前景，包括自动化织物缝合技术，同时，该展馆还在模仿海胆外形的设计方面表现出色。

建筑的施工优势以及管理同样会随着气候的变化以及环保意识的变化而变化。这也在一定程度上体现了木材的优势，正如之前所提到的那样，它具有极强的防腐性。[5] 出乎意料的是，木建筑导致火灾的可能性要低于20世纪传统的钢材和混凝土建筑。[6] 在进行推断时，2002年的一个案例分析建议在长远的环境背景下，人们要更倾向于使用木材。"与制作胶合板梁相比，我们要花费两倍到三倍多的精力，六到十二倍多的化石燃料去制作钢梁。"

近代，木材产品加工方面的技术进步带来了富有诗意和美感的前景，人们的态度随之改变，大量的木材产品都在以下的八个项目中得以体现。

Wood's renaissance in the construction industry over the past decade is strongly connected to refinements in the production of engineered timber, which has, in fact, existed since the early 20th century. Thus, in a relatively short space of time a suite of new timber products has transitioned from niche product to standard material, familiar to most builders and designers. Of this suite of engineered products, "cross-laminated timber (CLT), also known as glulam[1], is considered the new kid on the block."[2]

New computer technology has meant that engineered timber products can now be machined and manipulated into more and more complex forms, greater lengths and breadths, and more quickly and cheaply.[3] These developments coupled with wood's lightness and strength to weight ratio mean it is also well suited to prefabrication and roofing, and is increasingly being utilized in large-scale structures with confidence and innovation.[4] The project featured on the cover of this issue, the University of Stuttgart's ICD/ITKE Research Pavilion, demonstrates the extraordinary possibilities of the most cutting edge technologies, including robotic, textile-sewing techniques, while doing a remarkable job of emulating sea urchins.

Construction priorities and regulations have also shifted with growing climate change and environmental awareness, which partly account for timber's ascendency; for timber is, as mentioned, a superior sequesterer[5] and, counter intuitively,

ICD/ITKE研究馆，2015-16，斯图加特
ICD/ITKE Research Pavilion 2015-16, Stuttgart

位于芬兰的塞拉基乌斯博物馆的Gösta馆由MX_SI事务所设计，其设计理念主要是为了保留Joennimei庄园宝贵的遗产价值以及围绕在它周围的、美丽的、迷人的森林美景，并使二者相得益彰。

后建的这座令人瞩目的新木构建筑在内部空间与外部空间之间建立了一种流动的对话。内部是精心设计的、微微流动的体量。在这些大型玻璃和木框的背后是无处不在的风景。这些构件打破了现有建筑体量的整体性。

用设计师的话说："从外面看来，建筑物是由一系列垂直的木框架构成的，这些框架遵循且强调了内部构造的韵律。在这些框架之间，一个垂直立面通风系统由单独捆在一起的云杉木条制成，以打破材料的结构限制。"

木材的使用促进了建筑和景观的一体化进程。MX_SI事务所的设计方案荣获了2013年的西班牙国际建筑奖（国际类）。

这座建筑的基础、室外覆层以及室内饰面都是采用木材制成的，是芬兰较为著名的、几乎全部采用木材来建造的大型公共建筑物之一。

位于法国Rillieux-la-Pape、由Tectniques建筑师事务所设计的Hacine Cherifi体育馆内设体育馆、多功能运动馆以及行政区，共占地2500m²。

建筑师是这样形容结构体系的："从内到外，结构体系是由覆盖在板条上的三层云杉板、定向刨花板（构成箱式基座）、36cm高的箱式框架（采用稻草来填充）、40mm厚的木纤维保温板、防雨棚以及与当地的花旗松木瓦相接的三层花旗松木板覆盖层构成的。"

在大量混合使用木材和混凝土的结构中，金色的木覆层成为这一

wood constructions can actually present a lesser fire risk than typical steel and concrete structures of the 20th century.[6] A 2002 case study recommended wood on further environmental grounds when concluding that "... *it takes two to three times more energy and six to twelve times more fossil fuels to manufacture steel beams than it does to manufacture glulam beams.*"[7] Attitude changes along with the discovery of new poetic and aesthetic possibilities have accompanied the most recent technical advancements in timber production, a number of which are showcased in these eight projects.

MX_SI's Serlachius Museum's Gösta Pavilion in Finland was designed to compliment and respect the high heritage value, Joennimei Manor, and the beautiful, petrified forest that surrounds it.

The subsequent, striking new timber structure establishes a fluid dialogue between interior and exterior spaces and contains carefully considered, subtly shifting volumes. The landscape is omnipresent behind generous glazing and timber mullions, which both serve to break up the architectural volume.

In the architects' words: *"On the outside, the building presents a series of vertical mullions that follow and emphasize the rhythm of the interior structure. Between the mullions a ventilated facade system was designed of spruce wood strips twisted independently to the tectonic limit of the material."*

The use of wood facilitates the integration of architecture with landscape. MX_SI's design was recognised with the Spanish International Architecture Award 2013 (International category).

The building, with its timber foundations, external cladding and internal finishes, is also noteworthy as one of the first large-scale public edifices made almost completely from wood in Finland.

The Tectoniques Architects-designed Hacine Cherifi gym-

Hacine Cherifi体育馆，法国
Hacine Cherifi Gymnasium, France

庞大木建筑的主体，且该结构置于浅色的、填充稻草的、用麦秆做成的木材框架之上。

建筑所体现的质朴以及由内到外的通透性，主要是通过自始至终都坚持使用一种材料——木材，以及暴露在外的结构美感来实现的。特别是材料，旨在平衡建筑物所展现的壮观规模。

风格独特的箱式结构因它极富表现力的天花板结构而变得与众不同。天花板设有凸起的天窗。透过天窗，充足的自然光能够照射进来，并且无眩光。这些天窗体现了这座乐高式建筑的简易质朴。

尽管在某些方面，与其他七个项目相比，Hacine Cherifi体育馆有些华而不实，但它的坚固性、充足的阳光、特定的功能性以及些许娱乐性（通过定制的红色场馆设备得以体现），都让人念念不忘。

相反，位于巴黎的、极具纪念意义的Créteil教堂的二次开发主要集中在复原建筑物以及建造一座和原教堂同等体量的建筑方面。

两座大型壳体建筑，或者说是木质船体结构，结合在一起，构成了主教堂，位于圣坛上方，像是祈祷中的双手。一座独立的、细长的木尖塔立在旁边，三个三角形洞口内设置了三个古钟。

在木材专家Sylva Conseil的指导下，"船形外壳"采用层压木材覆盖，插入云杉曲面板和花旗松木板条，花旗松木板预先遮挡起来，以确保老化的速度均匀统一。

据建筑师所说："云杉木建成的拱形结构的重复性，赋予了建筑内部以韵律，彰显了这座船形建筑的装饰风格，使这一神圣地方具有一定的密集性。"至于他们所采用的主材料："木材是一种非常天然、充满生命力的材料，简洁，庄重，这也使它十分适合建筑的曲线设计。它自身的温度使其形成具有友好氛围的社区风格……"

nasium, in Rillieux-la-Pape, France, encompasses gymnastics and multi-purpose sports halls and an administration space totaling a vast 2,500m² area.

The architects explain the structural system as follows, *"From the inside out, the system is composed of three-ply spruce panels laid over the battens, the OSB panels, which make up the base of the boxes, the 36cm deep box framework filled with straw, a 40-mm insulating woodfibre panel, the rain barrier and the cladding made up of three-ply Douglas Fir boards onto which timber tiles, also made of locally-sourced Douglas Fir, are fixed."*

This golden timber unequivocally dominates in this vast mixed wood and concrete structure that is set upon a light, straw-filled timber baton frame.

The building's simplicity and inside-out transparency come from the consistent use of one material, timber, and exposed structural aesthetic. The former in particular came from the intention to subdue the building's imposing scale.

The resulting stylized box is distinguished by an expressed ceiling structure whose raised skylights provide ample natural light without glare and, together, suggest the simplicity of a young child's Lego construction.

While the Hacine Cherifi Gymnasion is slicker, in some respects, than the other seven projects, solidity, ample light, and stripped-back functionality, with a dash of playfulness – via customised red gym equipment – are the take-home impressions of this complex.

In contrast, the monumental Notre-Dame de Créteil Cathedral redevelopment in Paris, was focused on reinvigorating architecture and doubling the Cathedral's congregation.

Two giant shells, or wooden hulls, form the main Cathedral as they draw together, above the altar, like hands almost sealed in prayer. A detached slender, timber spire stands nearby, with three triangular apertures exposing the same number of ancient bells.

Knarvik社区教堂，挪威
Community Church Knarvik, Norway

Créteil教堂的扩建项目，法国
Créteil Cathedral Expansion, France

　　在该项目以及本系列的其他项目中，光线是非常重要的，它经过了精心的策划，产生独特的氛围。彩绘玻璃安装在建筑船形结构南部的最前面，突出了建筑的神圣感，同时也使建筑全天都能够接收到日光。

　　Reiulf Ramstad事务所设计的Knarvik社区教堂，是一个非常有吸引力且经过全方位精心设计的建筑。它的设计理念非常注重气候、环境、建筑的地位（作为一处朝圣的场所），并考虑了其他社区庆祝以及表演活动。

　　这一引人注意且在一定程度上考虑了未来设计的建筑，被人们使用优美的词语描绘为"折叠的星星"[8]。该建筑的灵感来自于当地的山峰和峡湾。屋顶是尖尖的，不容易积雪。结构的长立面交替使用了垂直的木板和玻璃，使周围的景观能够融入建筑的肌理中，并且始终凸显建筑的广阔性。

　　尽管这一现代设计并没有显著的宗教标识，但是由于其被社区居民广泛地使用，这座建筑物仍然具有标志性，主要是因为其设有瘦削陡峭的尖塔。从远处看，教堂立面木材和玻璃的交替使用给人一种传统石廊的感觉。

　　Knarvik教堂和巴黎圣母院是不同的。Knarvik教堂的平静感和神圣感主要来自于其广阔性、多变性以及与周围建筑物的巧妙结合。反之，巴黎圣母院之所以具有强烈的氛围感，主要是因为其建筑的封闭性、统一性以及密集性。而Knarvik教堂之所以能实现这两种效果，主要是因为使用木材作为外围护结构以及木材之间采用了合适的间距的缘故。

　　苏黎世动物园新建的大象馆与Créteil大教堂有两点共同之处：创

Under the guidance of wood specialist, Sylva Conseil, the "hulls" were clad in laminated wood, stuck bows of Spruce and Douglas Fir strips, the latter pre-shaded to ensure uniform ageing.

According to the architects, the "..*repetition of spruce arches gives rhythm to the interior, a decorative style of the hulls. The intention here is … to characterise the density of a sacramental space*".

And of their principle material: "*Wood is a natural, living material, at once humble and noble. It lends itself perfectly to the design of the building's curves. Its warmth also serves as a pattern of a fraternal community,…*"

Light is an important, well-orchestrated and atmospheric feature in this project, and most others in this collection: Here stained glass is positioned at the head of the southern hull, enhancing the sense of this being a sacred space and allowing for all-day sunlight.

The Community Church Knarvik by Reiulf Ramstad Arkitekter is a dramatic and refined building-in-the-round, which has been strongly informed by climate, context and the building's role as a place for worship, and other community celebrations and performances.

This dramatic, in some ways futuristic, structure – which has been beautifully described as a folded star[8] – was inspired by the mountains and fjords in the region: Snow would slide easily from these sharp roof plains; the structure's long elevations have vertical bands of glass alternating with timber that admits the surrounding landscape into the fabric of the building, all the while enhancing the architecture's expansive quality. While this contemporary design has avoided overt religious iconography, due to broad community use, the building's function is still signaled through its splinter-steep spire, and the alternation of glass and timber along the church's elevation could also evoke traditional colonnades from afar.

曼彻斯特的新麦姬癌症中心，英国
New Maggie's Cancer Center in Manchester, UK

新的木结构和与墙体穿插结合的曲形木屋顶。

大象馆与众不同的特色是其龟壳状的网格结构木屋顶，为预制的工程木壳体结构。木壳体的洞口在场地经过切割和移动能够营造出一种迷宫的感觉。

前面提到的工程木的前景以及与之有关的绘制和预制加工技术在这里都得到了充分的体现。在这些措施出现以前，这种如此壮观的定制曲形横截面以及长度是很难付诸实践的，而且经济花费也较为巨大。

在设计位于英国英格兰曼彻斯特的木构麦姬癌症中心时，福斯特及其合伙人建筑事务所并没有将传统的医疗机构设计作为参考。相反，他们采用了给人以温暖氛围的天然木材，具有触感的肌理和表面，并且在中心位置设置了厨房，设计了透明的玻璃温室以及网格式屋顶结构，这些设计使花园填补了所有的透明缝隙，营造一种充满欢迎氛围的、如家的充满阳光的空间，如同避难所，让人们感到舒适自在。

最后这两个建筑项目所体现的特点和舒适性以及主材料和周围树木的结合，似乎彰显了木材是极具生命力的事物，而这一点主要是由木材与自然和四季之间的种种联系来体现的。⁹

光线、平静以及与建筑物的结合

一个项目建筑师说了这样一段话，他的话从不同角度都适用于该系列的大部分建筑："在一定程度上，人们是推崇将外部空间穿插进建筑物之中的。"其他穿梭于建筑中的元素是照明与多变性，无论是从体量还是氛围上来说，都取决于材料的选择和处理。

麦姬癌症中心采用轻质木材建成的网格结构木屋顶，以及大象馆

The Knarvik Church is distinguished from the Parisian Cathedral, in that its sense of serenity and sacredness comes from an expansiveness, levity and connection to the landscape, whereas the Cathedral gathers its potent atmosphere from a sense of enclosure, uniformity and density, although both effects of the church have been achieved with the enveloping and spacing of timber.
The Zürich Zoo's new Elephant House shares two things in common with the Créteil Cathedral: An innovative wooden construction, and curved timber roof, which morphs into walls.
The Elephant House is distinctive for its tortoise shell-like lattice timber roof, made from a prefabricated engineered wood shell, which had its openings cut and removed on site to create a maze-like effect.
The aforementioned new possibilities of engineered wood, and its associated mapping and prefabrication technologies, are evident here. For prior to their existence, such ambitiously customized, large curved cross-sections and lengths would not have been practically or economically viable.
Foster and Partners determinedly eschewed any design reference to traditional health institutions when designing the timber-framed Maggie's Cancer Center in Manchester, England. Instead, the warmth of natural timber, tactile fabrics and surfaces, the placement of a kitchen at the design center, and use of transparent glasshouse and lattice roof structures, allowed the garden to occupy all transparent gaps, and create a welcoming, homey, light-filled space, that provides refuge and comfort.
The character and comfort found in these final two projects and their connection between primary material and surrounding foliage seem to affirm and project the notion of wood as something alive, due to its association with nature and the seasons.⁹

Sognefjellshytta高山酒店的新入口，挪威
Sognefjellshytta High Mountain Hotel's New Entrance, Norway

墙体和屋面边缘切割的天窗网格，用一种质朴和不拘小节的方式将这些建筑与自然界连接在一起，并且有意识地利用自然光来渗透建筑，以打破建筑的厚重感。

进一步说，自然光、绿色植物以及麦姬癌症中心的花园景色通过以下几种方式得到了增强：移动门使中心可以直面大花园。每间治疗室前面都有一处狭小的绿色空间。建筑物的南面对着温室休闲区开放[10]。癌症中心结构和材料的轻盈性、照明以及氛围都充分地呈现，且梁还是人们识别空间变化的标志。据建筑师所言，"他们的设计从视觉上将建筑融入周围的花园中"，营造出一种与众不同的、引人注目的环境。

同样地，Gösta馆的切口、反射玻璃以及平行小路，形成了与周围景观紧密结合的门或者林间小道。

教堂和新社区建筑无墙的设计、切口、木框以及柱廊使建筑富于变化。同时，也使这些建筑能够与周围景观相融合，或追随景观的踪迹，把体量分割成若干个小部分"[11]。

天然材料——木材将单个体量包围起来，也因此优化了建筑由内到外的整体效果。建筑从外部到内部使用木材来进行无缝拼接，并且对灯光进行了巧妙的处理，其内的空间呈现出一种静谧的感觉，使整座建筑都被贴近自然的大地色材料所包裹。Sewn木材研究馆采取的也是相似的设计准则，力求达到相似的效果，只是采取了一种更为激进的方式。

旧与新

传统的柱廊或者竖框（或由木材制成，而非石头或者混凝土，或由玻璃焊接），是新与旧之间、或者是永恒的元素和建筑语言之间的相互

Light, calm ness and connection to landscape

The words of one project architect could easily apply, in various ways, to most of the buildings in this collection: *"external spaces, in a sense, are encouraged to penetrate inside the building"*. The other strong thread running through these projects is lightness or levity – of volume and atmosphere – emanating from material choice and its treatment.

The lightweight timber roof lattice structure of Maggie's Cancer Center and cutting edge skylight webbing of the Elephant House roof-walls, connect these structures in a more rustic or informal way to the natural world, while intentionally saturating them with natural light and breaking up the architectural density.

Further, natural light, greenery and garden views in Maggie's Cancer Center have been enhanced in the following ways: "sliding doors open the center to a large garden, each treatment room faces a small green space and the building's south side opens to a greenhouse retreat space"[10]. The Cancer Center's lightness of structure and material, illumination and atmosphere have also been achieved with beams signaling room changes, *"visually dissolving the architecture into the surrounding gardens"*, according to the architect, to create a distinct and inviting environment.

Similarly, architectural "incisions", reflective glass and parallel paths in the Gösta Pavilion, all create "doors or forest walkways" into the surrounding landscape.

Levity has variously been achieved through wall-less volumes, incisions, timber mullions and colonnades in the Church and the Social House. This has allowed these buildings to merge into or trace the landscape, while encouraging volumes to "decompose into smaller fragments"[11].

The use of the natural material, timber, has made it viable to wrap singular volumes, thereby, optimising the inside-out effect. The seamless use of timber from exterior to interior,

1. http://www.timber.net.au/index.php/timber-wood-products-glulam.html,
"a type of structural engineered wood product comprising a number of layers of dimensioned timber bonded together with durable, structural adhesives that are moisture-resistant."
2. http://blogs.aecom.com/connectedcities/the-renaissance-of-timber-structures/
"The use of this type of timber (CLT) has increased in volume by over 600 percent in Europe in the last decade, going from being a niche building material to a standard form of construction, familiar to most builders and designers."
3. http://blogs.aecom.com/connectedcities/the-renaissance-of-timber-structures/
4. http://www.theengineer.co.uk/issues/sept-2012-online/timber-renaissance/
http://blogs.aecom.com/connectedcities/the-renaissance-of-timber-structures/
5. http://www.seattle.gov/dpd/Blog/Wood%20Concrete%20of%2021st%20Century.pf
"Research suggests that wood, a local resource, vastly outperforms other common building materials like concrete and steel in terms of both carbon emissions and sequestration and engineered timber is equivalent..."
6. http://blogs.aecom.com/connectedcities/the-renaissance-of-timber-structures/
"There are also concerns around fire safety on large scale timber buildings, with a perception that they're at more risk of burning down. In reality, however, the use of thick timber sections actually has far more inherent fire resistance than an equivalent steel structure."
7. Sandin Peters Svanström, *Life cycle assessment of construction materials: the influence of assumptions in end-of-life modelling. International Journal of Life Cycle Assessment 19*, p.723-731, 2014
8. https://www.yatzer.com/knarvik-church-reiulf-ramstad-arkitekter
9. Silvio Carta + Stefano Tesotti, "The Future is Wood", C3 #297
10. Foster + Partners project text 11. MX_SI Project project text
12. Silvio Carta + Stefano Tesotti, "The Future is Wood", C3 #297

作用的典范：小教堂和大教堂采取的是富有戏剧性的现代化的演绎，但仍保留着传统的尖塔。从远处看，这些尖塔向人们宣告了自身作为一处朝圣之所的庄严感，在传统的哥特式宗教建筑传统中，这种设计仿佛能够连接天堂。通过使用木材竖框，博物馆和社区建筑将传统的市政建筑的柱廊与木材结合在一起。麦姬癌症中心和大象馆也融入了不同的永恒感——自然界永恒存在的、普遍的元素。

这些项目背后的建筑师通过使用木材来反映当地环境的一些方面。每个项目也都展示了材料的一个卓越特性，主要体现在其与光线的结合，低调的时尚展示，以及未采用与其他材料相对比的方式。

尽管这篇文章开始的主题是来自世界的、令人瞩目的大型建筑物的木表皮的集合，但这些建筑的精髓却是贯彻在其内部的巧妙设计以及丰富的表面材料。实际上，由于木材极高的灵敏性和创新性，木材所带来的暖意和"诗意"（在其他文章中也有所讨论）[12] 得到了进一步探讨。

along with the deft consideration of light, has also enabled the creation of calm spaces that are enveloped in an earthy material reference to nature. The Sewn Wood Research Pavilion embodies, in a more radical way perhaps, similar principles to similar effect.

Old and new

Traditional colonnades or mullions – made from wood rather than stone or concrete or conjured through vertical insertions of glass – are examples of the interplay between new and traditional or timeless elements and architectural languages: Dramatic, modern interpretations of church and cathedral, nonetheless, retain steeples, which announce them from afar as places of worship, as they stretch to the heavens in the tradition of the gothic and religious architecture; the timber mullions of Museum and Social House conjure colonnades of traditional civic buildings with wood; Maggie's Cancer Center and the Elephant House blend with a different sense of the timeless – the constant and universal elements of nature.

The architects behind each of these projects have reflected something of the local environment by using wood sourced from it; each also demonstrates a great respect for the material, by allowing it to shine, in its understated-fashion, without rivalry from other materials.

While this article's starting theme was the timber skins of a collection of striking large-scale buildings from around the globe, the sophistication of design and richness of surface material have been carried through, in the majority of these projects, to their interiors. Indeed, the warmth and "poetic possibilities" of wood, discussed elsewhere,[12] have been explored with great sensitivity and innovation. Susanne Kennedy

ICD/ITKE研究馆2015-16
ICD + ITKE + University of Stuttgart

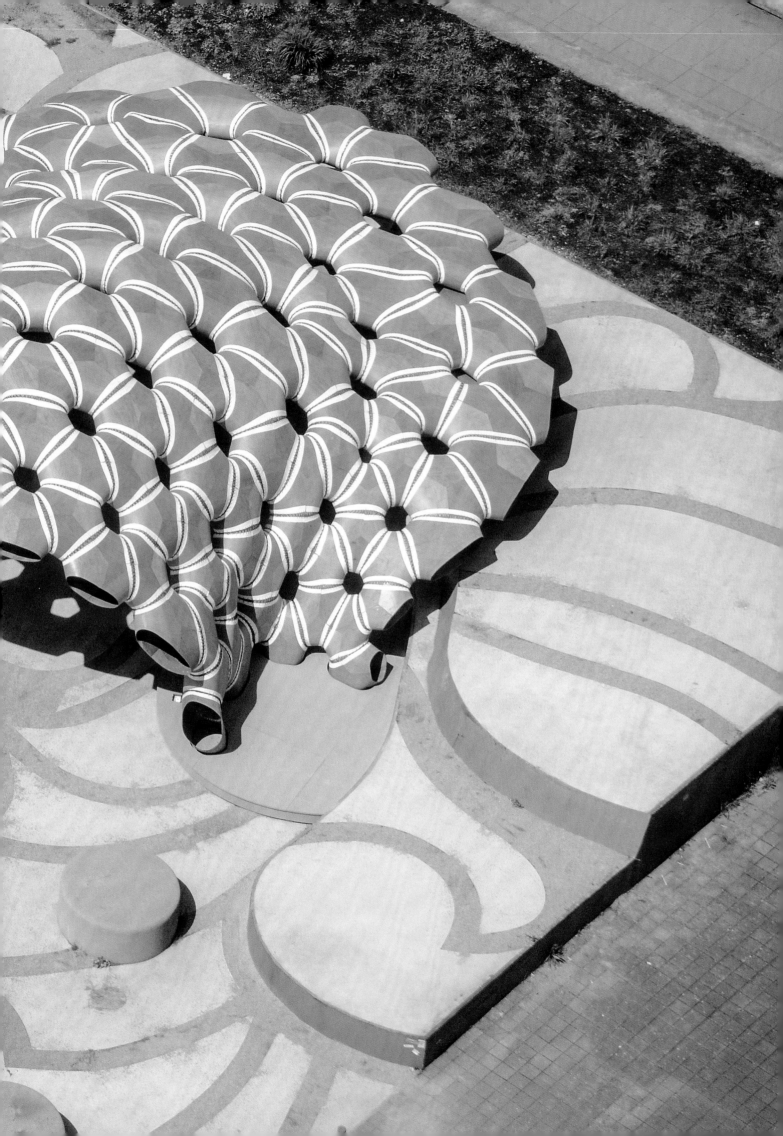

斯图加特大学的计算机设计学院(ICD)和结构建造与结构设计学院(ITKE)共同设计和建造了一座全新的研究馆。该研究馆向人们展示了针对分段式木构外壳所使用的自动化织物加工技术。它是全世界范围内第一座将木构件进行工业化缝合的建筑,是一系列成功建成的研究馆之一,该系列的研究馆都向人们展示了在建筑领域计算机设计、模拟以及加工过程的发展潜力。该项目是由学生、来自多学科建筑队伍的建筑师、工程师、生物学家以及古生物学家组成的研究人员设计和施工的。

壳体建筑的仿生研究

ICD/ITKE研究馆2015-16的设计特点是采用双重自下而上的策略。这一策略的制定主要基于对分段式天然板材结构的仿生研究以及缝合胶合板薄层所使用的新型自动化加工技术。该项目的设计思路始于对海胆的结构形态分析。同时,加工技术的发展也使一种新产品得以开发出来,即定制的、可自动化缝合的山毛榉层压胶合板制成的双层分段式构件,可弹性弯曲。在这个木结构的施工过程中,织物连接技术使分段式木构外壳能够以轻巧的形式来连接,同时也非常具有表现力。

利用木材的材料属性和结构理念

基于生物学原理以及物质特点,这座研究馆的材料系统被设计为一个双层结构,如同楯形目海胆第二次生长时的内部结构。建筑构件主要包含极薄的木条。利用木材的各向异性,这些木条可定制为层压木条,不同纹理方向和不同厚度的木板具有不同的硬度,可形成拥有不同半径的构件。这些原始的平板木条可弹性弯曲,从而形成特定的形状。在这一变形的过程中,利用自动化缝合技术,木构件都可固定为特定的形状。这样一来,151个形状不同的构件便加工完成。当这些构件组装在一起时,一个十分坚固的双层曲形壳体结构就此形成。

一座融合建筑学、工程学以及生物学的先驱建筑

这座研究馆一共有151个构件,都采用自动化缝合技术来加工,且每一个构件都由三层独立的山毛榉胶合层压木条制成,直径在0.5m到1.5m之间。构件的特殊形状和材料构成经过编程,以适应特定的结构和几何外形需求。这个项目采用的织物连接技术使建筑不再需要使用金属紧固件。整个结构重780kg,占地85m²,跨度达9.3m。材料的厚度和跨度的平均比为1:1000,每平方米的重量是7.85kg。

整体的设计与大学校园里特定的场地条件相呼应。它创造了一个半室外的空间,使地面与之融为一体,如同由座位组成的景观,同时这处空间还面向附近的公共广场开放。这个项目通过形成更复杂的空间布局(而非一个简单的外壳)来体现这个研发系统在形态方面的适应性,该研究馆向人们展示了利用计算机将生物学原理进行的整合以及

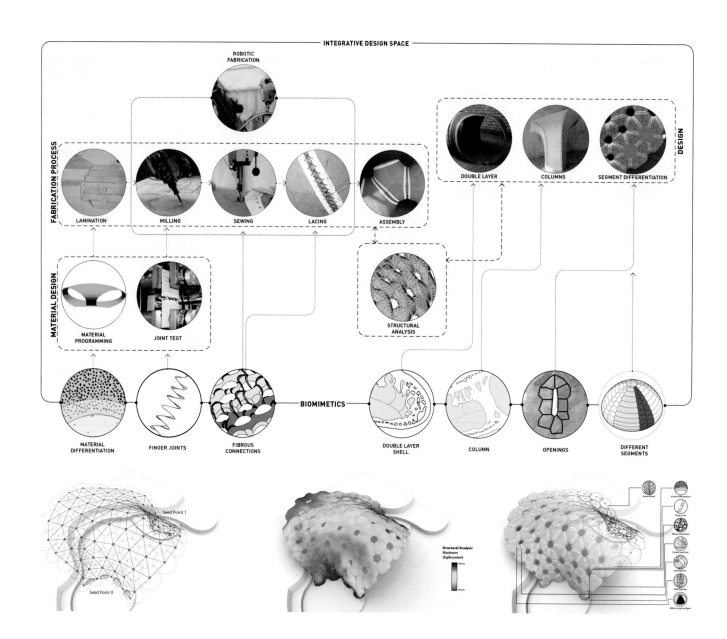

材料、外形和自动化加工之间的复杂相互作用是如何引导新型木结构施工方法产生的。这一多学科的研究方法不仅推动了具有表现力的、节省材料的轻质建筑的出现，同时，它还探索了创新的空间特点，拓展了木建筑构造的可能性范围。

ICD/ITKE Research Pavilion 2015-16

The Institute for Computational Design (ICD) and the Institute of Building Structures and Structural Design (ITKE) of the University of Stuttgart have completed a new research pavilion demonstrating robotic textile fabrication techniques for segmented timber shells. The pavilion is the first of its kind to employ industrial sewing of wood elements on an architectural scale. It is part of a successful series of research pavilions which showcase the potential of computational design, simulation and fabrication processes in architecture. The project was designed and realized by students and researchers within a multi-disciplinary team of architects, engineers, biologists, and palaeontologists.

Biomimetic Investigation into Shell Structures

The development of the ICD/ITKE Research Pavilion 2015-16 is characterised by a twofold bottom-up design strategy based on the biomimetic investigation of natural segmented plate structures and novel robotic fabrication methods for sewing thin layers of plywood. The project commenced with the analysis of the constructional morphology of sand dollars. At the same time, a fabrication technique was developed that enables the production of elastically bent, double-layered segments made from custom-laminated, robotically sewn beech plywood. Introducing textile connection methods in timber construction enables extremely lightweight and performative segmented timber shells.

Employing the Material and Structural Logic of Wood

Based on both the biological principles as well as the material characteristics, the material system was developed as a double-layered structure similar to the secondary growth

in sand dollars. The building elements consist of extremely thin wood strips. Instrumentalising the anisotropy of wood, these strips are custom-laminated so that the grain direction and thickness corresponds with the differentiated stiffness required to form parts with varying radii. Thus, the initially planar strips can be elastically bent to find the specific shape. In this deformed state, the elements are locked in shape by robotic sewing. In this way, 151 geometrically different elements could be produced, which result in a stiff doubly curved shell structure when assembled.

A Demonstrator on the Intersection of Architecture, Engineering and Biology

The pavilion consists of 151 segments that were prefabricated by robotic sewing. Each of them is made out of three individually laminated beech plywood strips. Ranging between 0.5 and 1.5 m in diameter, their specific shapes and material make-up are programmed to fit local structural and geometrical requirements. The textile connections developed for this project allow overcoming the need for any metal fasteners. The entire structure weighs 780 kg while covering an area of 85 m² and spanning 9.3 m. With a resulting material thickness / span ratio of 1/1000 on average, the building has a structural weight of only 7.85 kg/m².

The overall design responds to site-specific conditions on the university campus. It establishes a semiexterior space that integrates the ground topography as a seating landscape and opens towards the adjacent public square. At the same time it demonstrates the morphologic adaptability of the developed system by generating more complex spatial arrangements than a simple shell structure. The research pavilion shows how the computational synthesis of biological principles and the complex reciprocities between material, form and robotic fabrication can lead to innovative timber construction methods. This multidisciplinary research approach does not only lead to performative and material efficient lightweight structure, it also explores novel spatial qualities and expands the tectonic possibilities of wood architecture.

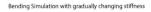

Bending Simulation with gradually changing stiffness

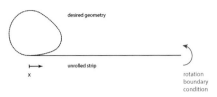

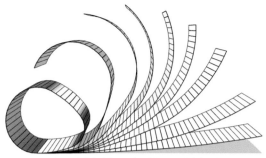

有限元素的非线性几何外形分析，可大跨度地移动
Geometric non-linear Finite Element Analysis with large displacements

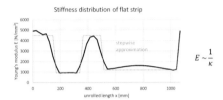

$$\kappa = \frac{1}{R} = \frac{M}{EI}$$

$$E \sim \frac{1}{\kappa}$$

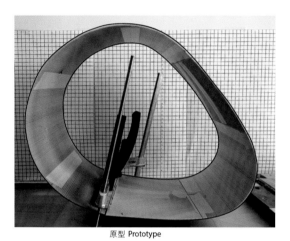

原型 Prototype

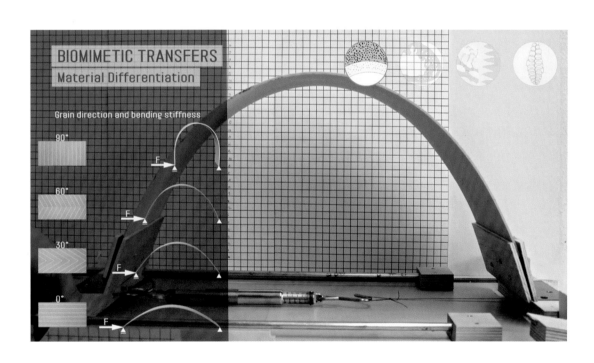

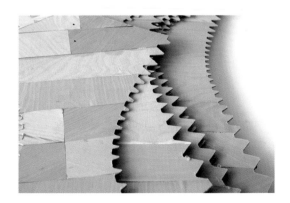

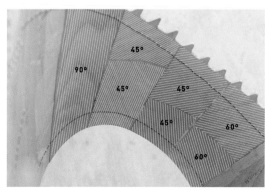

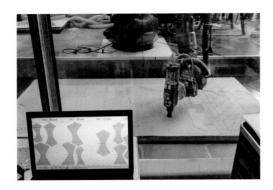
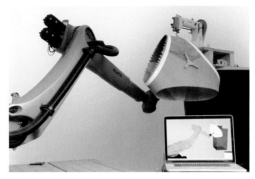
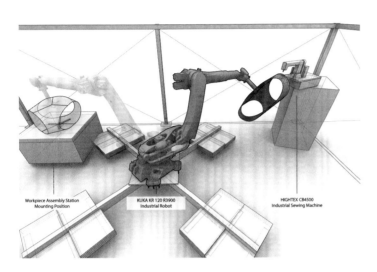

项目名称：ICD/ITKE Research Pavilion 2015-16
地点：Keplerstr. 11-17, 70174 Stuttgart
ICD设计：Prof. Achim Menges
ITKE设计：Prof. Jan Knippers
科学理念开发：Simon Bechert, Oliver David Krieg, Tobias Schwinn, Daniel Sonntag
理念开发、系统开发，加工和施工人员：
Martin Alvarez, Jan Brütting, Sean Campbell, Mariia Chumak, Hojoong Chung, Joshua Few, Eliane Herter, Rebecca Jaroszewski, Ting-Chun Kao, Dongil Kim, Kuan-Ting Lai, Seojoo Lee, Riccardo Manitta, Erik Martinez, Artyom Maxim, Masih Imani Nia, Andres Obregon, Luigi Olivieri, Thu Nguyen Phuoc, Giuseppe Pultrone, Jasmin Sadegh, Jenny Shen, Michael Sveiven, Julian Wengzinek, Alexander Wolkow(with the support of Long Nguyen, Michael Preisack, and Lauren Vasey)
参与人员：Prof. Oliver Betz_Department of Evolutionary Biology of Invertebrates, Prof. James Nebelsick_Department of Palaeontology of Invertebrates, University of Tuebingen
有效楼层面积：85m² / 外壳面积：105m² / 构件数量：151
规模：11.5×9.5 m / 竣工时间：2016.4
摄影师：©courtesy of ICD-ITKE University Stuttgart

曼彻斯特的新麦姬癌症中心

Foster + Partners

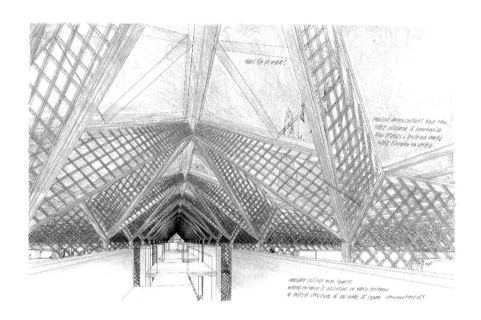

西南立面 south-west elevation

新麦姬癌症中心根据曼彻斯特的克里斯蒂医院的要求而建，是康沃尔郡公爵设立的。新中心将会有更长远的慈善目标：为全英国的癌症病人提供一种自由且实际的情感和社交支持。

麦姬中心遍布英国及其海外，提供了一处充满欢迎氛围的、"不在家但胜似家"的住所。这是一个避难所，被癌症折磨的人们能够在这里找到情感和物质支撑。由于被麦姬·凯瑟克·詹克斯所提出的建设一种新型护理中心的蓝图所激励，麦姬中心十分重视建筑设计所带来的力量，希望能以此激励人们的精神，在他们的治疗过程中提供帮助。在构造园区的同时，曼彻斯特中心的设计意图是在一处花园背景内营造一种如家的氛围，人们位于两边树木成荫的道路尽头即可一眼瞥见它，距离克里斯蒂医院及其肿瘤科仅几步的距离。

该建筑选取的地点阳光充足，整座建筑只有一层。既保持了低矮的轮廓，又反映了周围街景的住宅规模。屋顶通过三角形的采光天窗进行采光，且由轻巧的木格梁进行支撑。这些梁柱将内部的空间自然地分成不同的区域，在视觉上使建筑和周围的花园融为一体。中心将个人的私密空间与图书馆、健身房以及聚会的茶餐厅这些多样化空间相结合。建筑的中心是厨房，厨房的中心放置了一张大型公共餐桌。建筑内没有设置一些具有参考作用的构件，如走廊和医院的标示，以营造家庭式的空间。为此，材料采用了暖色系的天然木材以及具有触感的表面。屋顶高高地耸立在建筑中心，形成一个夹层。这样一来，后勤办公室都被安置在夹层，厕所和储藏室则位于其下层，以在视觉上将这座建筑物的各个部分自然地连接起来。

环顾中心，其设计重点主要集中在自然光线、绿化区以及花园视野方面。景观庭院突出了建筑的线性平面，整个西立面延伸至宽敞的阳台，且阳台被深悬挑的屋顶所遮蔽，以防止雨水渗透。滑动玻璃门使建筑面向花园开放。花园是由丹·皮尔森工作室设计的。

位于东立面一侧的每间治疗室和咨询室都可面向其私人花园开放。建筑的南端向前延伸，以将一间温室围合起来。温室是一处充满光线和自然的地方，提供了一处花园式休闲空间，也是一个人们聚会、手工劳动地方，人们在此可享受自然与户外美景带来的治疗效果。这里也是一处种植了各种花卉和其他农产品的空间，这些植物能在某时某刻病人感觉最脆弱的时刻，给予他们一定的目标。

该中心由Foster+Partners事务所的设计师和工程师负责，其富有特色的定制家具则由Norman Foster和Mike Holland设计，他们在该事务所中带领团队负责工业设计，设计了厨房构件、桌子、餐具柜以及其他架子等构件。

New Maggie's Cancer Center in Manchester

The new Maggie's Cancer Center in the grounds of The Christie Hospital in Manchester was opened by The Duchess of Cornwall. The new center will further the charity's aim to provide free practical, emotional and social support to cancer patients across the UK.

Located across Britain and abroad, Maggie's Center are conceived to provide a welcoming "home away from home" – a place of refuge where people affected by cancer can find emotional and practical support. Inspired by the blueprint for a new type of care set out by Maggie Keswick Jencks, they place great value upon the power of architecture to lift the spirits and help in the process of therapy. The design of the Manchester center aims to establish a domestic atmosphere in a garden setting and, appropriately, is first glimpsed at the end of a tree-lined street, a short walk from the Christie Hospital and its leading oncology unit.

The building occupies a sunny site and is arranged over a single storey, keeping its profile low and reflecting the residential scale of the surrounding streets. The roof is naturally illuminated by triangular roof lights and it is supported by lightweight timber lattice beams. The beams act as natural

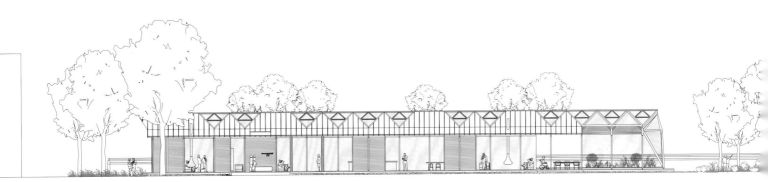

西北立面 north-west elevation

项目名称：Maggie's at the Robert Parfett Building / 地点：Manchester, United Kingdom / 建筑师：Foster + Partners
项目团队：Norman Foster, David Nelson, Spencer de Grey, Stefan Behling, Darron Haylock, Diego Alejandro, Teixeira Seisedos, Xavier De Kestelier, Mike Holland, Richard Maddock, Daniel Piker, Elisa Honkanen
结构工程师：Roger Ridsdill Smith, Andrea Soligon, Karl Micallef, Mateusz Bloch / 环境工程师：Piers Heath, Evangelos Giouvanos, Nathan Millar
消防工程师：Thouria Istephan, Michael Woodrow / 估料师：Gardiner＆Theobald / 景观顾问：Dan Pearson Studio
照明顾问：Cundall / 规划师：IBI Taylor Young / 认可检察员：AIS (Approved Inspector Services) / 温室设计顾问：Fleur de Lys
主要承包商：Sir Robert McAlpine / 木结构设计和建筑承包商：Blumer Lehmann / 甲方：Maggie's
摄影师：©Nigel Young (courtesy of the architect)

用地面积：1,922m² / 有效楼层面积：500m² / 建筑高度：6.15m / 建筑长度：52.65m / 建筑宽度：19.25m / 建筑规模：one story above ground ＋ mezzanine
结构：timber lattice structure, shallow foundation with concrete pads / 材料：timber lattice beams and pillars, cross laminated timber roof panels, timber cladding, bronze roof, brick floor, aluminium sliding windows / 设计时间：2013 / 施工时间：2014 / 竣工时间：2016

partitions between different internal areas, visually dissolving the architecture into the surrounding gardens. The center combines a variety of spaces, from intimate private niches to a library, exercise rooms and places to gather and share a cup of tea. The heart of the building is the kitchen, which is centered on a large, communal table. Institutional references, such as corridors and hospital signs have been banished in favour of home-like spaces. To that end the materials palette combines warm, natural wood and tactile surfaces. The roof rises in the center to create a mezzanine level, where support offices are placed on a mezzanine level positioning on top of toilets and storage spaces, maintaining natural visual connections across the building.

Throughout the center, there is a focus on natural light, greenery and garden views. The rectilinear plan is punctuated by landscaped courtyards and the entire western elevation extends into a wide veranda, which is sheltered from the rain by the deep overhang of the roof. Sliding glass doors open the building up to a garden setting created by Dan Pearson Studio.

Each treatment and counselling room on the eastern facade faces its own private garden. The south end of the building, extends to embrace a greenhouse – a celebration of light and nature – which provides a garden retreat, a space for people to gather, to work with their hands and enjoy the therapeutic qualities of nature and the outdoors. It will be a space to grow flowers and other productions that can be used at the center giving the patients a sense of purpose at a time when they may feel at their most vulnerable.

The center, designed and engineered by Foster + Partners, also features bespoke furniture designed by Norman Foster and Mike Holland who head out the industrial design team in the practice. These include kitchen units and table, sideboards and other shelving units.

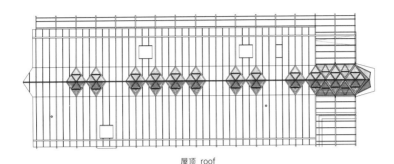
屋顶 roof

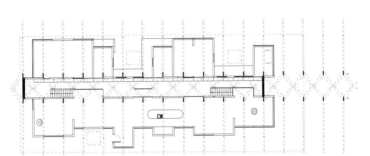
夹层 mezzanine floor

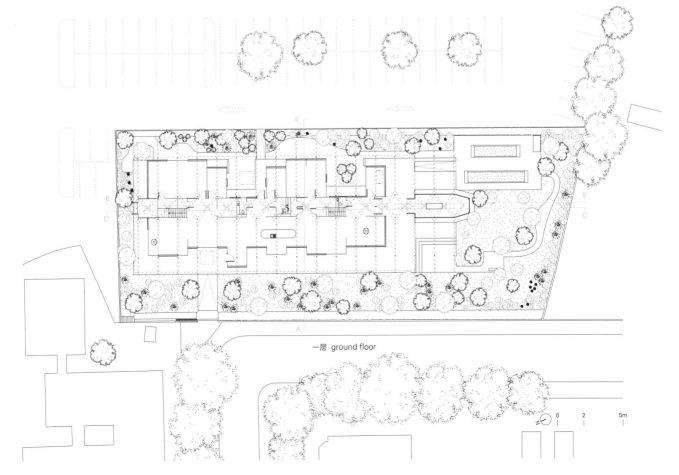
一层 ground floor

A-A' 剖面图 section A-A'

B-B' 剖面图 section B-B'

C-C' 剖面图 section C-C'

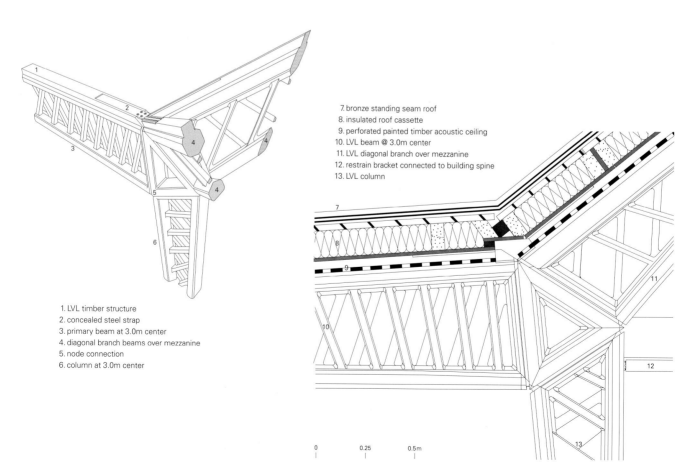

1. LVL timber structure
2. concealed steel strap
3. primary beam at 3.0m center
4. diagonal branch beams over mezzanine
5. node connection
6. column at 3.0m center
7. bronze standing seam roof
8. insulated roof cassette
9. perforated painted timber acoustic ceiling
10. LVL beam @ 3.0m center
11. LVL diagonal branch over mezzanine
12. restrain bracket connected to building spine
13. LVL column

圆形木亭

Christiansen Andersen

建筑师：Christiansen Andersen
地点：Gammel Kongevej 91B, 1850 Frederiksberg C, Denmark
主要建筑师：Mikkel Kjærgrd Christiansen, Jesper Kort Andersen
资助人：Dreyer Foundation, MOELVEN Denmark, Copenhagen Technical College, and the Danish Architects' Association in collaboration with the Agency for Palaces and Cultural Properties
总建筑面积：275m²
施工时间：2015 (6 weeks in the workshop, 2 weeks in King's Gardens)
摄影师：©Hampus Berndtson (courtesy of the architect)

圆形木亭是一座临时性建筑。当你参观哥本哈根最著名的公园之一——国王的花园时，圆形木亭会给你耳目一新的感觉。这座公园是一处最特别的空间，是城市中的一片绿洲，也是一处人们主导的地方，而在这里，文化遗产正在消亡。为什么我们要去国王的花园呢？如果是一个人去，我们可能会想要到公园放松一下，欣赏一下蓝天，感受一下树木带来的静谧与自由。如果是两个人，我们可能想要体验一下二人世界。最后一点，也是非常重要的一点，当一群人来这儿的时候，主要是社交原因。以上三点是圆形木亭吸引人们的主要原因。

我们如何在国王的花园中增建丰富游客体验的空间呢？我们希望能够打造一处这样的空间：当人们坐在树下的长椅上时，能够最大限度地浏览到周围的景色。这样一来，这处空间就成为圆形木亭的室外部分。受国王的花园里那条对称的小路的启发，我们还想建造一条小路。最后，我们想创造一种意境，让人们能够站在蓝天下的草坪上享受自由呼吸。怎么样才能创造一个与周围环境相得益彰的结构呢？为此我们近距离地观察花园里的各个元素，观察其线条、比例以及空间。通过研究，我们找到了问题的答案以及合适的策略。面向亭子内角的四个区域呈低矮的姿态，并为人们提供了一段入口长台阶。当进入圆亭

以后，人们仍然可以瞥见外面的花园美景，但同时也被限制在这一特定的区域。

该亭子的中心区域是一片圆形草丛，用来玩耍、野餐和举行重要事件，此外，人们还可以在这里做其他想做的事情。一条有顶的人行漫步道使人们可以在此坐下休憩及散步，孩子可以在此奔跑。

公园里的树木与透过木结构穿过来的光线，树林、青草以及夏天空气的气味共同形成了一种韵律，创造了与众不同的空间体验。

圆亭采用的材料都是Moelven公司生产的北欧可持续性木材，由哥本哈根技术学院的木匠实习生建造。建筑使用的是尺寸为36mm×68mm的挪威标准板条。屋顶覆层使用的是尺寸为18mm×68mm的板条，是通过将板条对半切割实现的。主结构采用的是三加二板条模式，这也是我们建造柱和梁所用的基本材料。隔墙是用尺寸为36mm×68mm的完整木条以及尺寸为36mm×34mm的对半切割的木条共同建造而成的。圆亭的各个部分都是春季实习生在工作间预备和预制的，并且在两周的时间内将其安置在公园中。

Around Pavilion

The pavilion is a temporary work of architecture adding a fresh perspective to your visit to one of the finest parks in Copenhagen, the King's Gardens. Here, cultural heritage meets evanescence in a most particular space, an urban oasis in the midst of the city, a place where the guests in the park rule. Why do we go to the Gardens? As one person, we probably do it to rest and enjoy the blue sky, the quietness and the freedom of trees. As two, we probably do it to share an intimacy. And last, but not least, as group we go there for social reasons. These three phenomenons became the key to the way the pavilion was drawn.

How can we apply new spaces that enrich the experience of being in the Gardens? We wanted to create a space where the

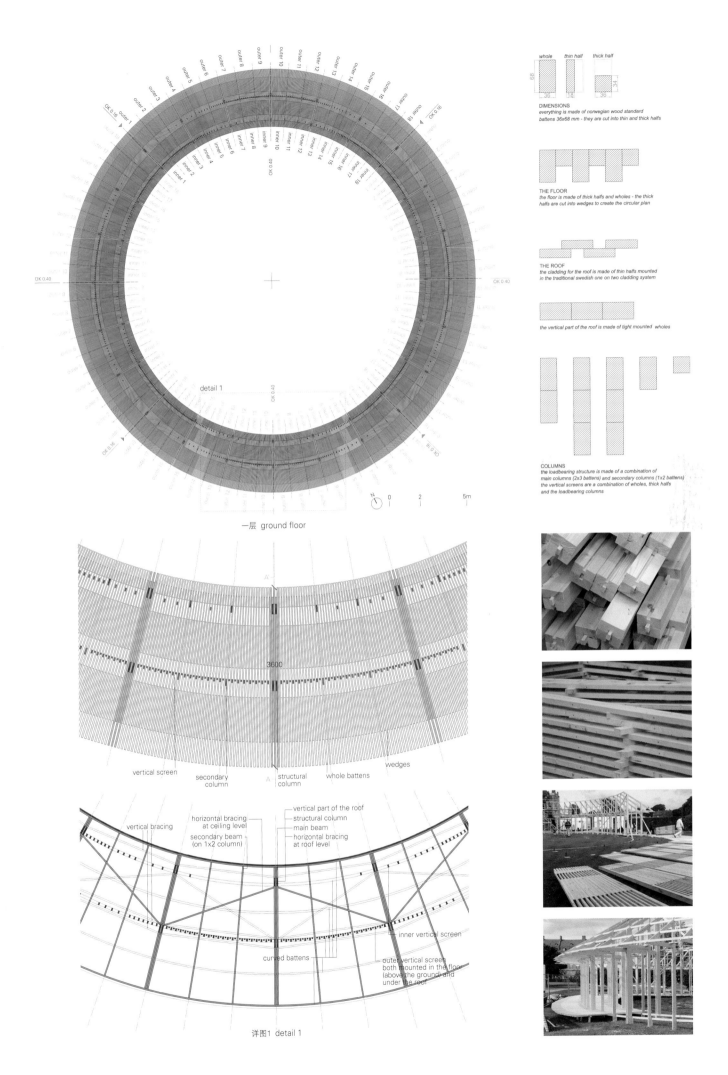

horizon of the Gardens was experienced at a long sight under the tree branches. This became the outer part of the pavilion. And we wanted to create an alley inspired by the symmetrical alley in the Gardens. And finally we wanted to create the ultimate experience of standing in the grass under the sky. How can we create an object that adapts beautifully the surroundings? By looking very closely at the elements in the Gardens, at the lines, the proportions and the spaces. Through these studies we found the answers and a way make it fit.

Four places facing the corners of the pavilion bow down, providing a long step for entering. When inside the pavilion, one can still see glimpses of the garden outside, but is at the same time capsulated in a remarkable space.

In the middle there is a circle of grass for play, picnic, events and whatever one feels like and a covered "promenade" provides opportunities for seating and walking, or running that the kids do.

The trees in the garden create a rhythm with the light

glimpses through the wooden structures of the pavilion and the scent of wood, grass and summer air unites in a spatial experience like no other.

The pavilion structure is constructed of sustainable, Nordic wood from Moelven and built by carpenter's apprentices from the Copenhagen Technical College. It's constructed by using only standard Norwegian battens with the dimensions 36 x 68 mm. By cutting the battens in half the roof cladding with the dimension 18 x 68 mm was created. The main structure consists of two and three battens, which allowed us to create columns and beams, and the separating walls are created by single and half battens with dimensions 36 x 68 mm and 36 x 34 mm. In this way all the parts for the pavilion were prepared and prefabricated by the apprentices in the workshop during the spring and joint in the park in just two weeks.

Christiansen Andersen

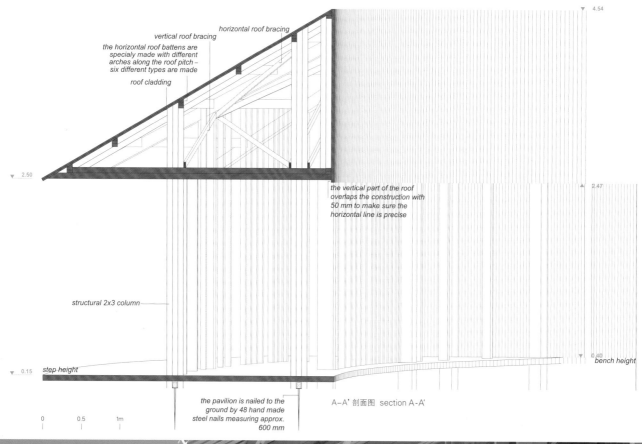

A-A' 剖面图 section A-A'

木材——诗意与实用 Wood – Poetic and Practical Possibilities

Caltron的新型社区
Mirko Franzoso Architetto

这一项目斩获了设计比赛的奖项，该比赛面向35岁以下的设计师开放，赋予了意大利北部克莱斯小镇的Caltron的新型社区一定的活力。该地点是当地居民聚会的地方，孩子、年轻人、成年人以及老人可以在此见面、讨论，同时产生各类生活经历交织在一起所形成的社区归属感。社区的功能配置也是以此目的为出发点，既是为了孩子也是为了年长的人，面向所有人开放。

这座新建筑被认为是具有历史特色的小镇和乡村景观之间的纽带，同时也限制了城镇的扩张及肆意化的土地滥用。这座建筑的所有施工方面的选择都是基于其所扮演的重要社会角色及其所处的特殊环境而做出的。

这个新体量是由简单的建筑（以一种自然的方式嵌入地面），以及支撑建筑的大型基座构成的。建筑体量十分紧凑，且与周围的环境规模紧紧呼应，以保证这座小镇城市化进程的连续性。现代化风格的建筑是景观的一部分，并且与苹果园的背景融合，向人们彰显了这一地区那令人熟悉的、原有的存在感。立面以及壁柱间的虚实形成的韵律感使建筑在不改变色调及其他元素的情况下，能够延续其与苹果园之间的连续性。建筑整体都是用木材建造的。南面和北面立面都是用落叶松垂直板条覆盖的，而东面和西面则用落叶松柱条来定义。

窗户做后退设置，并且被立面保护起来，以形成内部和外部空间关系的连续性，同时也是为了减弱阳光对内部空间造成的影响。地下空间主要是围绕着歌剧院设计：其长形内嵌式彩色单体冲洗混凝土体量将停车场、操场以及社区建筑包围起来。

冲洗混凝土表面使用的当地产的斑岩使墙壁的颜色和周围环境很好地融合在一起。

建筑内宽敞的木质壁龛空间充满了欢迎的氛围，来迎接进入建筑内的人们。此外，它还减弱了长形墙壁所产生的威严感。室内空间的布局是按照公共管理功能来设定的。

建筑物的入口位于会议厅的上方，会议厅是一处十分空旷的空间，可以根据需要进行调整和改造。人们从会议室的北边望去，透过三道拉门，可以到达所有带暖气的房间，分别是浴室、厨房和储藏间。两处主要的空间，也就是厨房和会议室，都设有大型窗户，且窗户形成了与室外的连接性。

上楼以后，你将会抵达一个大型有顶的露台，露台是完全开放的，可自由出入。人们在露台上可以以一个更高的角度（而非街道水平）来观看山谷。支撑屋顶的木柱发挥了遮阳的作用，并且增加了空间的隐私性。在西面街道上方一侧还有另一条通往露台的通道。

综上所述，Caltron社区建筑旨在为整个社区提供一个重要的聚会点，并且还提供一个全新的观景视角，同时人们在此还能远眺整个城镇，是一个吸睛场所。

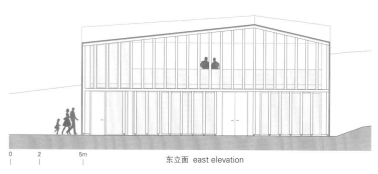

东立面 east elevation

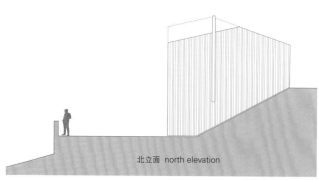

北立面 north elevation

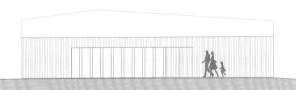

西立面 west elevation

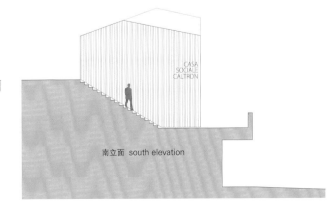

南立面 south elevation

New Social House in Caltron

The project that won the invited competition open to designers under 35 gave birth to the new social house in Caltron in the small town of Cles, in the northern Italy. This place is a meeting point for the inhabitants, where children, youth, adults and seniors can meet, discuss and grow altogether a community identity where life stories intersect. The functions are designed with this aim, both for children and for elderly people, and are accessible to all .

The new building is a "trait d'union" between the historic town and the rural landscape, but at the same time arises as a limit to the expansion of the town and then to the indiscriminate abuse of land. All the architectural choices of this building are based on its important social role and on the particular context in which it occurs.

The new volume is a system made by a simple architecture, inserted in a natural way on the ground and a powerful base that sustains it. The building is compact and proportionate to the surrounding ensuring the continuity of the urbanization of the small town. The contemporary architecture becomes part of the landscape and blends with the apple orchards background, to become a familiar pre-existence for the people from that area. The facade, with its rhythm of solids and voids realized by pilasters continues the succession of apple trees without altering tones and matter. The building

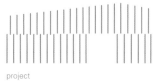

generator   project

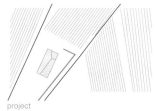

generator   project

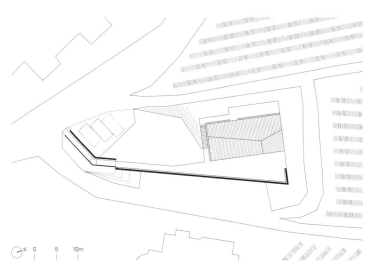

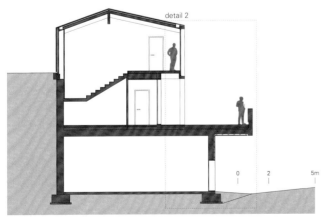

A-A' 剖面图 section A-A'

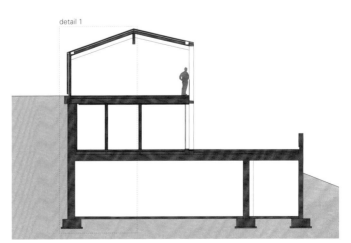

B-B' 剖面图 section B-B'

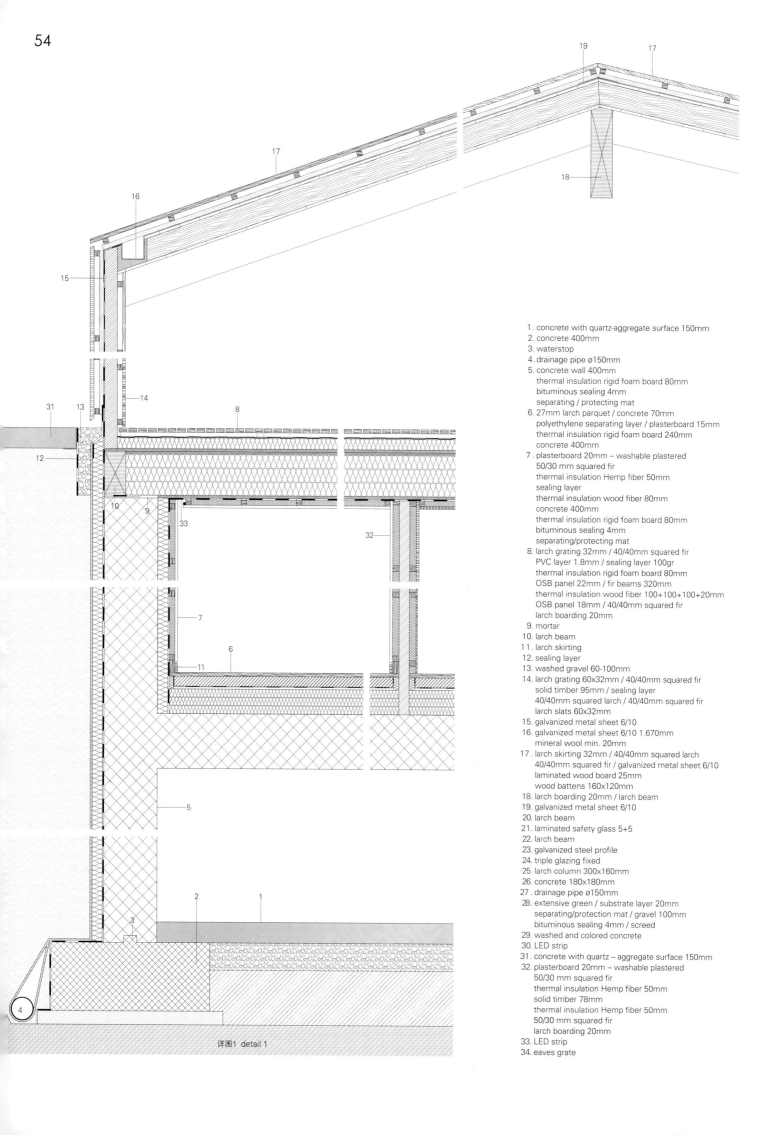

详图1 detail 1

1. concrete with quartz-aggregate surface 150mm
2. concrete 400mm
3. waterstop
4. drainage pipe ø150mm
5. concrete wall 400mm
   thermal insulation rigid foam board 80mm
   bituminous sealing 4mm
   separating / protecting mat
6. 27mm larch parquet / concrete 70mm
   polyethylene separating layer / plasterboard 15mm
   thermal insulation rigid foam board 240mm
   concrete 400mm
7. plasterboard 20mm – washable plastered
   50/30 mm squared fir
   thermal insulation Hemp fiber 50mm
   sealing layer
   thermal insulation wood fiber 80mm
   concrete 400mm
   thermal insulation rigid foam board 80mm
   bituminous sealing 4mm
   separating/protecting mat
8. larch grating 32mm / 40/40mm squared fir
   PVC layer 1.8mm / sealing layer 100gr
   thermal insulation rigid foam board 80mm
   OSB panel 22mm / fir beams 320mm
   thermal insulation wood fiber 100+100+100+20mm
   OSB panel 18mm / 40/40mm squared fir
   larch boarding 20mm
9. mortar
10. larch beam
11. larch skirting
12. sealing layer
13. washed gravel 60-100mm
14. larch grating 60x32mm / 40/40mm squared fir
    solid timber 95mm / sealing layer
    40/40mm squared larch / 40/40mm squared fir
    larch slats 60x32mm
15. galvanized metal sheet 6/10
16. galvanized metal sheet 6/10 1.670mm
    mineral wool min. 20mm
17. larch skirting 32mm / 40/40mm squared larch
    40/40mm squared fir / galvanized metal sheet 6/10
    laminated wood board 25mm
    wood battens 160x120mm
18. larch boarding 20mm / larch beam
19. galvanized metal sheet 6/10
20. larch beam
21. laminated safety glass 5+5
22. larch beam
23. galvanized steel profile
24. triple glazing fixed
25. larch column 300x160mm
26. concrete 180x180mm
27. drainage pipe ø150mm
28. extensive green / substrate layer 20mm
    separating/protection mat / gravel 100mm
    bituminous sealing 4mm / screed
29. washed and colored concrete
30. LED strip
31. concrete with quartz – aggregate surface 150mm
32. plasterboard 20mm – washable plastered
    50/30 mm squared fir
    thermal insulation Hemp fiber 50mm
    solid timber 78mm
    thermal insulation Hemp fiber 50mm
    50/30 mm squared fir
    larch boarding 20mm
33. LED strip
34. eaves grate

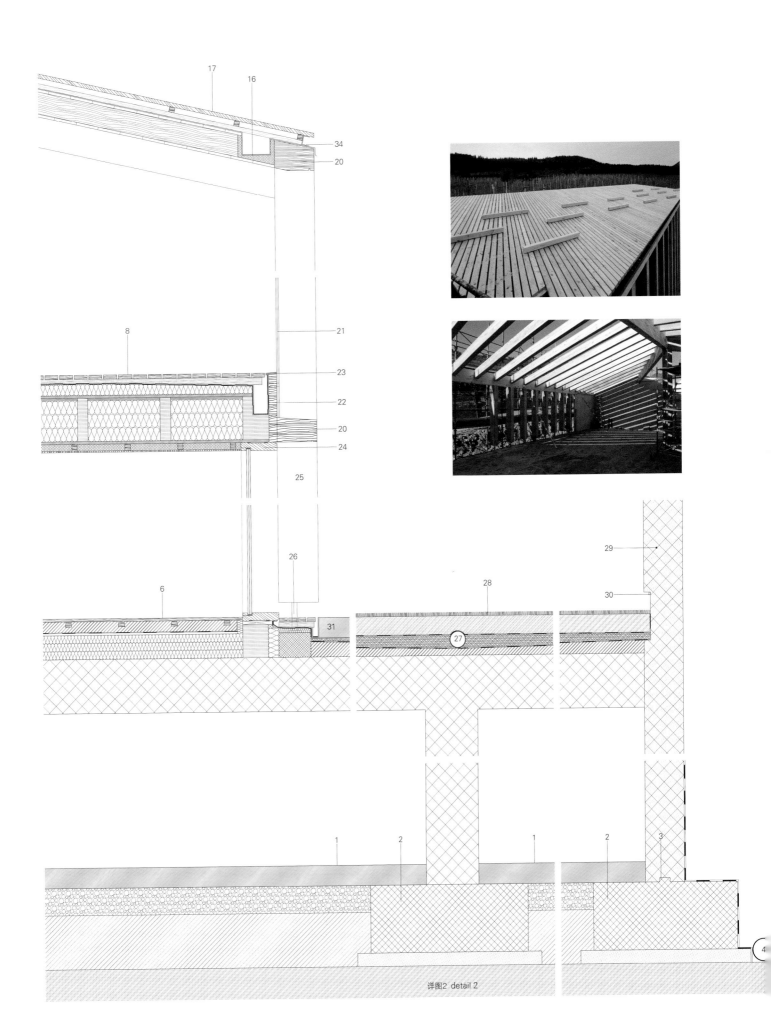

详图2 detail 2

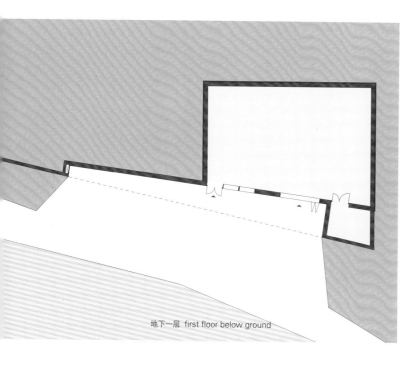
地下一层 first floor below ground

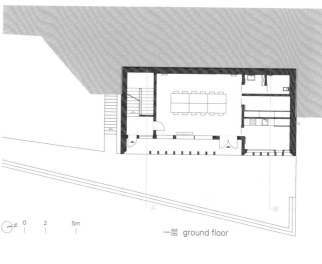
一层 ground floor

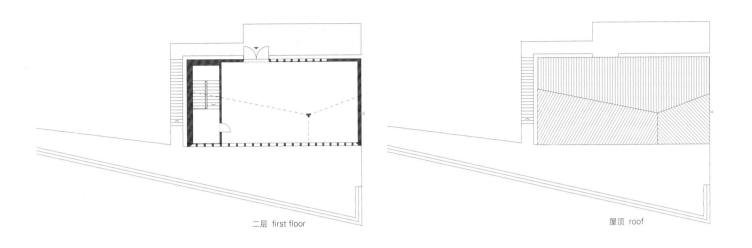

二层 first floor   屋顶 roof

项目名称：New Social House in Caltron / 地点：Cles-TN, Italy / 建筑师：Mirko Franzoso architetto / 结构工程师（混凝土）：ing. Paolo Leonardi, ing. Claudio Cristoforetti (Pro-engineering di) / 结构工程师（木材）：ing. Sergio Marinelli_Bolzano / 管道系统设计：p.ind. Walter Dallago / 电气系统设计：p.ind. Giorgio Rollandini / 建造商：F.lli Borghesi_wood, Mak costruzioni_concrete / 甲方：Comune di Cles / 封闭区域面积：440m² / 总面积：1,120m² / 造价：EUR750,000 / 设计时间：2012.10~2013.07 / 施工时间：2014.04~2015.08 / 摄影师：©Mariano Dallago (courtesy of the architect)

is entirely made of wood; both the north and south facades are covered with vertical slats of larch, while the eastern and western long fronts are defined by the vertical scan of the larch pillars.

The windows are set back and protected by the facade plane to generate a continuity of the relationship between interior and exterior, as well as mitigate the effect of the sun in the inner rooms. The underground volume wraps the entire opera: its long monolithic built-in washed and colored concrete that embraces the parking, the playground and the social house. The local porphyry emerging from washing concrete harmonizes the color of the wall with the surroundings.

The wide wooden niche that welcomes people who enter the building mitigates and lightens the authoritative presence of the long wall. The distribution of the interior spaces was defined by the public administration functional program. The entry is on the meeting room, a space completely empty, adaptable and modifiable according to needs. From the north side of the meeting room, you can reach through three sliding doors all the other heated rooms, i.e. respectively the bathroom, the kitchen and the storage. The two main spaces, the kitchen and meeting room enjoy the presence of large windows that provide continuity with the exterior.

Going upstairs you will arrive in a large covered terrace completely free and open. This part of the building allows a magnificence view of the valley from a higher point than the street level. The wooden pillars that support the roof act as sun blind and increase privacy within the space. Another access to the terrace is provided on the west upstream street. In conclusion, the Caltron social house, is aimed to be an important meeting point for the entire community and offers a new perspective on the landscape, but at the same time enjoys high visibility from the town and serves as attraction.

Hacine Cherifi体育馆
Tectoniques Architects

这座新体育馆是以小镇的拳击手Hacine Cherifi的名字命名的。1998年，Hacine Cherifi在阿斯宜贝拉改写了历史，成为继马塞尔·赛尔当之后第一个法国世界中量级拳击冠军。

Hacine Cherifi体育馆与Paul Chevallier学校综合体同属于一个开发项目，是Tectoniques事务所在Rillieux-La-Pape设计的另一项目。体育馆与带有褶皱状景观屋顶的Paul Chevallier学校相邻，其规模使自身看起来十分壮观，同时又给人一种安静的感觉。该建筑共有两个体育场馆，占地2500m^2，室内的高度分别为9m和12m。为了维护该机构的价值，建筑使用的是木材和混凝土混合的结构，没有过多的人工修饰，且这些木构件大都是预制的。

对公共空间做出的积极贡献

该设施包括一个体育馆（880m^2），一个多功能体育馆（1100m^2）以及一个能够容纳公众和运动员的区域。体量的布局比较简单。两个场馆呈直角的布局，可提供全面的服务。看台可容纳400多名看客。看台处提供了俯看多功能体育馆的视野，且看台后方还面向入口区以及自助餐厅开放，人们从外面的公共区便可看到入口区和自助餐厅。此外，人们通过道路一侧的全玻璃立面还可看见建筑内举行的活动，但不会对正在进行的活动或者训练造成干扰。

依环境而设计的比例

为了对周围环境造成最小的影响，建筑师充分利用了街道和场地内部的高度差，从而使部分建筑与斜坡融合。这样一来，入口便位于建筑物的上方，而不是直通体育馆的下方。因此，道路一侧的建筑高度降低了6m。这样的规模与临近学校以及附近住宅的规模相适应。

城市的过渡区

建筑物的外形十分简单，立面也较为朴素。学校建筑的线条是流畅的，经过了景观的美化，而较为单调和朴素的体育馆则恰恰相反，展现了其稳定性与牢固性。它占据了这个开发项目的北角，即Rillieux的两

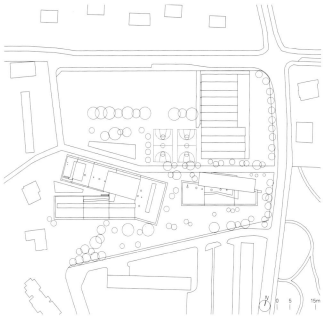

条主干道的十字路口处。Salignat大街一侧的东侧主立面通过其宏伟的、嵌入悬臂式体量的玻璃立面来清晰地向人们展示这座建筑物的公众地位。建筑物的后方,即朝西的一侧,特意营造出一种静谧的感觉,因住宅开发项目正处于建设之中,还未竣工。人们在朝南的一侧可俯看位于步行街的对面的学校。步行街为人们提供了通往学校的路口,同时还横穿整个开发项目。

木材和稻草

如同Tectoniques事务所的其他项目一样,这个项目的结构系统十分智能,也清晰明了。生混凝土墙覆盖了建筑的地下部分,人们位于室内时对其依稀可见。混凝土墙体的上方全部使用木材建造。建筑的主要结构是由一系列的门架构成的。门架由胶合层压格构式柱和梁组成,跨度长达34m,置于5.8m高的框架之上,为了通行方便,横梁分为两部分,在场地通过金属扣连接固定在一起。在施工过程中,采用抬升横梁并且将横梁安装到位的方法,一个引人注目的舞台就此形成。

外部壁板是由一系列的构件构成的,总面积为2000m²,这些构件预制成箱形,利用结构周围的外墙保温系统组装在一起。该系统由内到外是由压在板条上的三层云杉木板、箱形结构基座使用的定向刨花板、填充了稻草的36cm高的箱式框架、厚度为40mm的木纤维保温板、挡雨板以及覆层组成的。覆层由三层花旗杉板构成,本地产的花旗杉木制成的木砖固定在其上。

在建造比梁柱低或者高的屋顶时,横梁很好地发挥了作用。

该结构系统还包括大型玻璃横向板条,板条的位置面北,覆盖横梁的高度,这样一来既提供了充足的顶部照明,同时又避免了炫光。

红色的光芒打破了单一的色调

浅色调的使用主要是为了营造一种安静平和的氛围,同时也是为了使设施保持低调的姿态。地板是米黄色的,墙板使用浅色的木材。照明设备以及暖气片的位置都是经过精心布局的。唯一颇具色彩的便是红色的健身器材,与整体的单一色调形成了鲜明的对比。

Hacine Cherifi Gymnasium

The new gymnasium was named for Hacine Cherifi who is a boxer from the town. He made history in 1998 at the Astroballe when he became the first French world Middleweight champion since Marcel Cerdan.

The Hacine Cherifi gymnasium forms part of the same development as the Paul Chevallier schools complex, another Tectoniques project for the town of Rillieux-La-Pape. Next to the schools with their pleated, green roofs, its size makes it an imposing, yet silent, neighbour. The building houses two sports halls and support services over a total surface area of 2,500m² with internal ceiling heights of 9 and 12 metres. Faithful to the agency's values, the architecture uses a mixed wood/concrete structure without pretence or artifice, the wooden components being largely prefabricated.

An active contribution to the public space

The facility includes a gymnastics hall (880m²), a multi-purpose sports hall (1,100m²) and the spaces required to host the general public and sportspeople. The volumes are organised very simply. The two halls are set at right angles around a block which accommodates all the support services. The terraces offer seating for up to 400 spectators, which overlooks the multi-purpose sports hall and opens out at the back onto the entrance area and the cafeteria which are clearly visible from the public spaces outside. The fully-glazed facade on the road, offers glimpses of the activity inside the facility without disturbing the sports events or training in progress.

Scale adapted to the context

In order to limit the impact on the surrounding environment, the architects made use of the height difference between the street and the interior of the plot to integrate part of the building into the slope. This also allowed them to create at access at the top of the building (rather than the bottom which would lead directly into the sports halls). This made it possible to reduce the height of the building on the road by 6 metres. These dimensions match the scale of the neighbouring

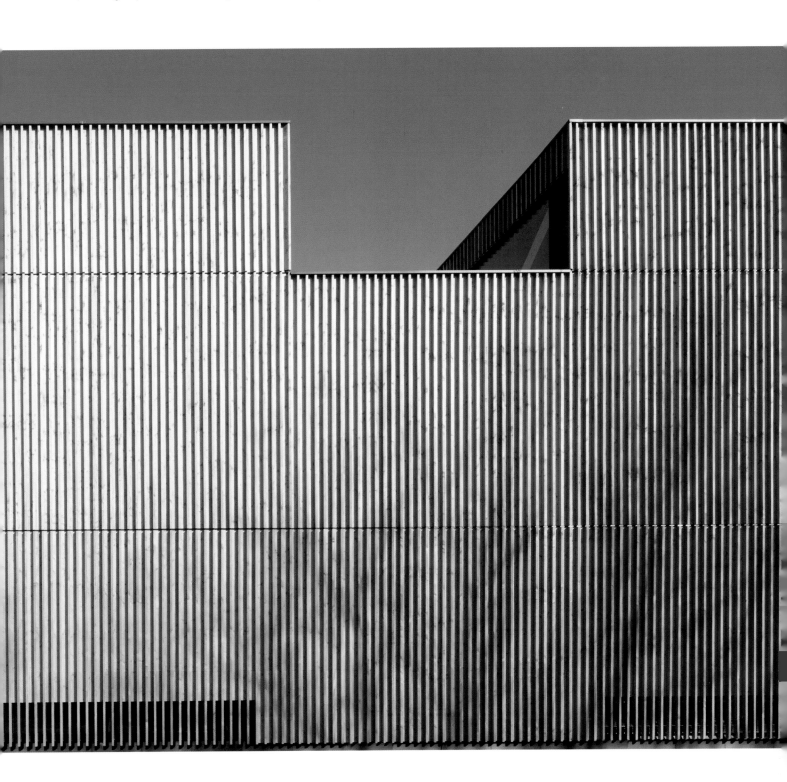

school and the residential fabric of the neighbourhood.

Urban mediator

The building has a very simple shape and the facades are unadorned. Whilst the style of the schools is fluid and landscaped, the simpler, more sober gymnasium, in contrast speaks of stability and solidity. It occupies the northern corner of the development at the crossroads between two of Rilleux's major roads. The main facade to the east on rue Salignat clearly states the facility's public status with a large, glazed facade, mounted with a cantilevered volume. At the rear to the west, the building is deliberately silent, as a residential programme is under construction and not yet finished. Towards the south it overlooks the schools, situated on the opposite side of a pedestrian alleyway which provides access to the schools and cuts across the development.

Wood and straw

As in all the Tectoniques agency's projects, the construction system is both intelligible and visible. The raw concrete walls which run the height of the underground section of the building are left visible from the interior. Above these walls, the structures are all made of timber. The main structure is formed from a series of portal frames. These are composed of glue laminated spaced columns and beams with a span of 34 metres, placed on a 5.8 metre framework. For the purposes of transportation, the beams were built in two parts, pinned and bolted together on site using metallic eyelets. Lifting and

东立面 east elevation

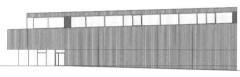
北立面 north elevation

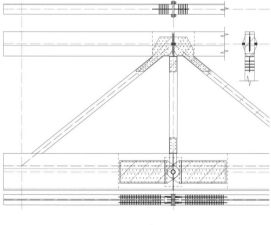

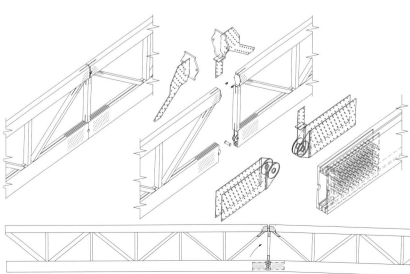

putting the beams into place made for a very dramatic stage in the construction process.

The 2,000 m² of components which make up the exterior siding are prefabricated into boxes, assembled using an external wall insulation system around the structure. From the inside out, the system is composed of three-ply spruce panels laid over the battens, the OSB panels which make up the base of the boxes, the 36cm deep box framework filled with straw, a 40mm insulating woodfibre panel, the rain barrier and the cladding made up of three-ply douglas fir boards onto which timber tiles, also made of locally-sourced douglas fir, are fixed. The beams are put to good use by constructing the roof above and below the height of the columns.

This system creates large, glazed and horizontal strips which face north and run the height of the beams. This provides abundant overhead lighting but avoids glare.

Flashes of red break through the monochromatic tone

Light shades have been used to create a calm, peaceful atmosphere and ensure the facilities remain discreet. The floor is beige, the wall panels are made of light-coloured timber and the position of the lights and radiators fit into a carefully designed layout. The only colour is the red gymnastics equipment which creates a stark contrast with the overall monochromatic tone.

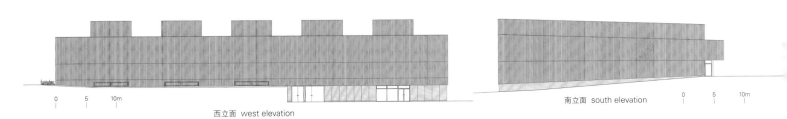

西立面 west elevation   南立面 south elevation

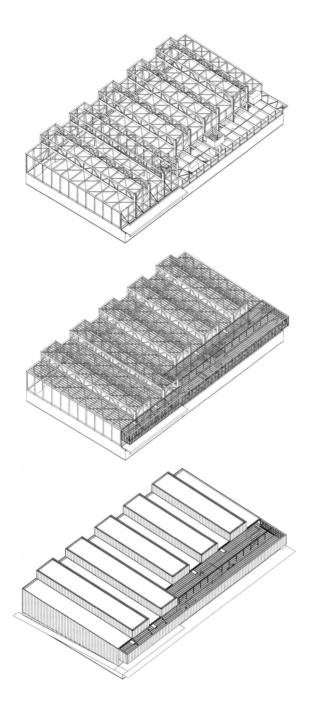

项目名称：Hacine Cherifi Gymnasium
地点：Rillieux-La-Pape, France
建筑师：Tectoniques Architects
合作：Arc / Wood structure _ Arborescence
保温和环境分析：Inddigo
混凝土结构和技术指导：Somival
经济顾问：Tectoniques Ingénieurs / 建筑控制：Veritas
健身和安保顾问：Socotec / 景观建筑师：Itinéraire Bis
甲方：Rillieux-La-Pape Town Council / 有效楼层面积：2,740m²
结构：raw concrete base(locally sourced), glue laminated timber framework(source: Germany), spruce columns and beams (source: France)
室外壁板设计：unfinished solid wood douglas fir timber tiles (locally sourced _ Vallée d'Azergue)
木工：External woodwork _ aluminium(source: France), Internal woodwork _ 3-ply spruce panels(source: Austria)
预算：EUR 3.7million ex. VAT
木材造价：EUR 1.15million ex. VAT
竣工时间：2015
摄影师：©11h45 et Tectoniques (courtesy of the architect)

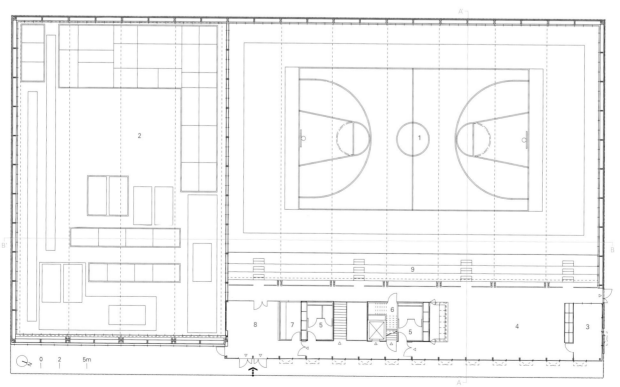

一层 first floor

1. 多功能体育馆	1. multi-sports hall
2. 体育馆	2. gymnastics hall
3. 俱乐部办公室	3. club office
4. 会客区	4. meeting area
5. 公共卫生间	5. public toilets
6. 垃圾投放区	6. dustbin location
7. 门房	7. watchman office
8. 入口	8. entry
9. 露天看台	9. bleachers

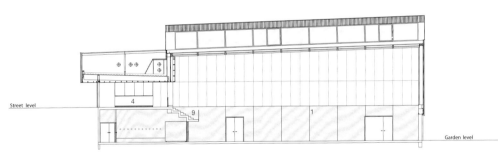

A-A' 剖面图 section A-A'

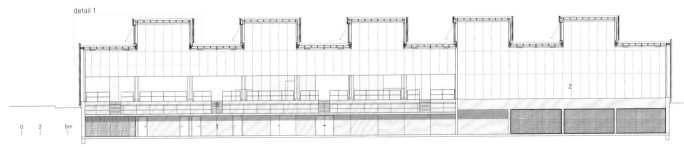

B-B' 剖面图 section B-B'

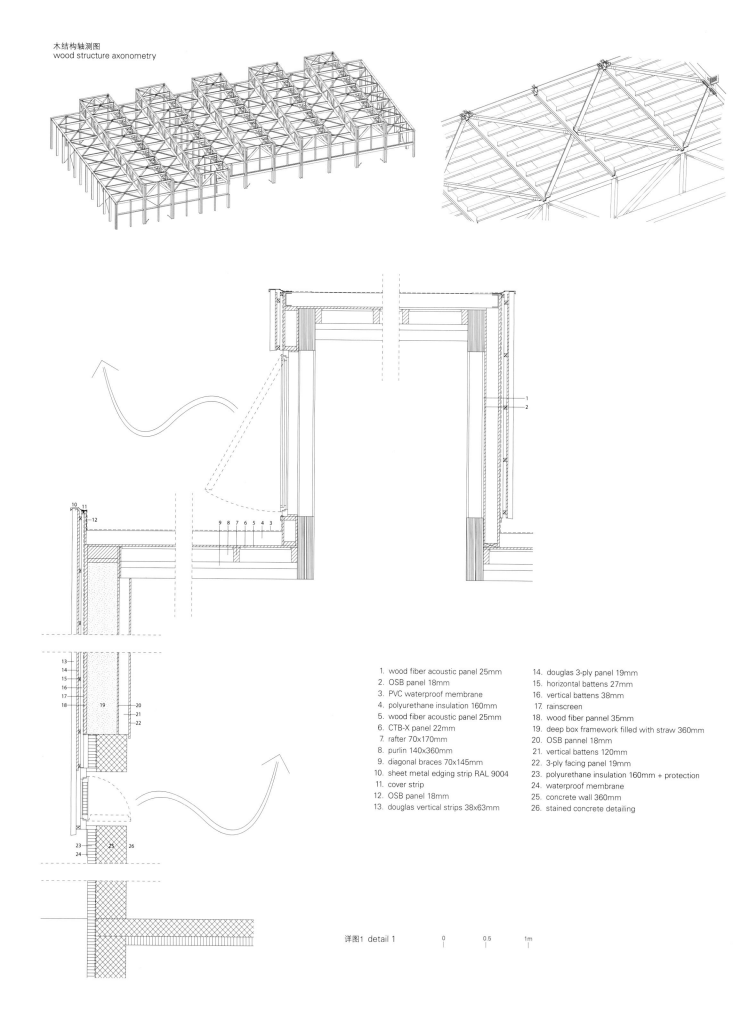

塞拉基乌斯博物馆扩建项目——Gösta馆
MX_SI

这一让人印象深刻而又与众不同的景观，在当代建筑的风格和历史遗迹之间建立了富有意义的对话。建筑方案对现存的特点，如主建筑、场地景观以及传统与现代文化给予了深刻的解读。设计的主要理念之一是对场地的正确解读进行回应。场地被看做一处绿地覆盖的高原，这里，现存的博物馆，即这一宏伟建筑的整体形象沿着景观轴线庄严地耸立着，并向Melasjärvi湖畔缓缓倾斜靠近。

设计策略包括建筑所处的场地要远离在视觉上隔断与现有场地特色和现存博物馆联系的区域。同时选址要充分利用能够眺望湖泊和后花园的视野优势。这一项目充分考虑了地形因素和方案所涉及的距离，以确保博物馆始终是该区域的主要建成结构。

新建筑的室内引进了室外的空间特点，这种做法主要是通过门廊（将室外与室内结合）来延伸入口广场完成的。新建筑的主结构是一个宽敞的门厅，与原建筑的一层位于同一水平线上，这一空间在视觉上将室内外连接起来，这一点主要是通过将室外景观空间切入建筑主体来实现的。为了呈现出展览布局的灵活性，建筑的结构成为立面的一部分，使整个空间拥有一个自由的平面。

这一项目的材料和外形使其呈现出一片抽象的、茂密的丛林形象。新建筑建在丛林中，实际上也是从概念上将场地的丛林以一种抽象的方式改造成的平行木框架。一方面，这些木框架定义了建筑整体的几何外形，同时，它们还对建筑进行横向渗透。其结果是从内到外结构框架都保持了平行的模式，从而组成了整座建筑。木材的使用也反映了当地工业化的历史以及塞拉基乌斯作为造纸生产地的传统。

Serlachius Museum Extention, Gösta Pavilion

A rich dialogue between contemporary architecture and heritage monument is mediated on an impressive and unique landscape. The architecture solution proposed a fine understanding on existing features such as main building, site landscape and traditional and contemporary culture.
One of the main concerns is to respond to a correct reading of the site. The site is understood as a green plateau where the existing museum, the manor's monolithic figure, stands imposingly along a landscape axis, sloping gently to the banks of Lake Melasjärvi.
The strategy consists in placing the building out of the zone that could block the visual relevance of the existing site features and the existing museum. Also the locations should take advantage of the privilege views to the lakes and treated gardens. The project considers the parameters of topography and distance to accommodate the program in order to allow that Joenniemi Manor house keeps being the dominant built

entrances
1. main entrance
2. secondary entrance
3. assembly hall / main entrance
4. service entrance (office/kitchen)
5. handling entrance
6. big 3x4m entrance directly to big hall

exterior areas/platos
7. park plateau
8. access yard
9. existing terrace
10. logistic area
11. access lane

buildings
12. Joenniemi Manor
13. new extension building
14. connection corridor
15. Autere's house
16. caretakers dwelling
17. cottage
18. shed
19. sauna
20. henhouse
21. garage
22. bridge

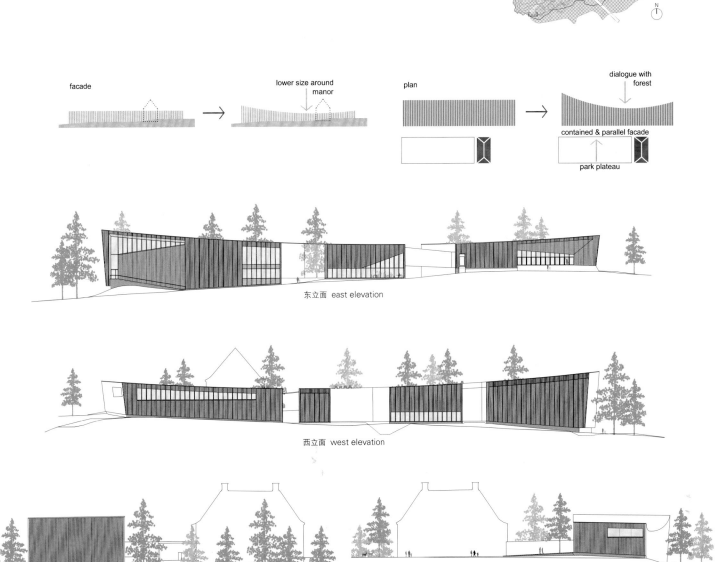

东立面 east elevation

西立面 west elevation

南立面 south elevation

北立面 north elevation

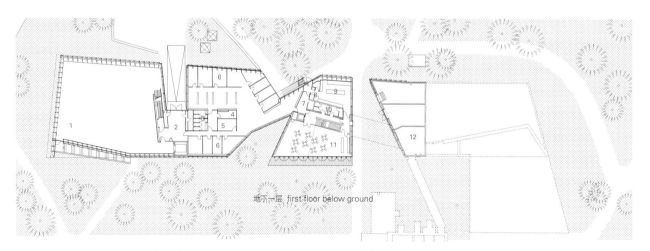

地下一层 first floor below ground

1.巡回展览区 2.操作室 3.卫生间 4.储藏室 5.小厨房 6.办公室 7.冷冻区 8.垃圾室 9.备餐间 10.餐厅卫生间 11.餐厅 12.衣帽间
1. travelling exhibition 2. handling 3. toliets 4. storeroom 5. kitchenette 6. office
7. freezer 8. waste room 9. preparation kitchen 10. restaurant toilet 11. restaurant 12. cloakroom

项目名称：Serlachius Museum Gösta Extension / 地点：Art Museum Gsta, Joenniementie 47, 35800 Mntt, Finland / 建筑师：MX_SI Architectural studio(Boris Bežan, Héctor Mendoza Ramirez, Mara Partida Muñoz) / 参与设计人员：Oscar Fabian Espinosa Servin, Olga Bomac, Elsa Bertran, Mariona Oliver, Jure Kolenc / 当地建筑师：Huttunen-Lipasti-Pakkanen Architects Oy / 项目管理：Pöyry CM Oy / 结构顾问：A-Insinöörit Oy / 电气工程顾问：Sähkötekniikka Kari Siren Oy / 景观建筑师：Maisemasuunnittelu Hemgård / 地热顾问：Ramboll Finland Oy / 甲方：Gösta Serlachius Art Foundation / 场地面积：80,000m² / 总建筑面积：3,500m² / 有效楼层面积：5,700m² / 桥体结构长度：53m / 造价：EUR19,500,000 / 设计时间：2011 / 施工时间：2012 /
竣工时间：2014 / 摄影师：©Pedro Pegenaute (courtesy of the architect)

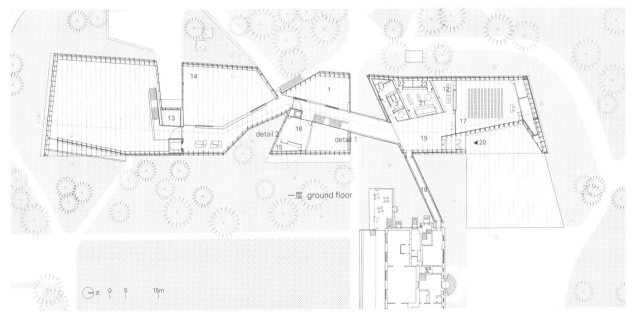

13.空中冬季花园 14.收藏区（画廊） 15.教学背景空间 16.博物馆教学设施 17.集会大厅 18.连接走廊 19.入门门厅+交流区 20.入口 21.售票区和商店 22.商店储藏室 23.控制室
13. hanging winter garden 14. collections (gellery) 15. pedagogy background space 16. museum pedagogy facilities 17. assembly hall 18. connecting corridor
19. entrance foyer+communications 20. entrance 21. ticket sales and shop 22. shop's storage 23. control room

structure of the area.

The outside spatial quality is brought inside the new building by extending the entry plaza through a porch that blends outside with inside. The new building is organized by a spacious foyer, placed at the same level of the ground floor of the existing mansion. This space obtains visual continuity between outside and inside by introducing incisions of landscape to the main building body. To allow flexibility of exhibitions layout, the structure of the building is part of the facade liberating the whole space as a free plan.

The project's materiality and geometry are presented as a densified abstract forest. The forest in the placement where the new building will be constructed is conceptually transformed in an abstract way of parallel wood frames. In one hand they define the overall geometry of the new building, but at the same time they also allow transversal permeability. The result is that the parallel pattern of the structural frames is maintained from outside and inside structuring the whole building. The use of wood is a reference to the local industry's history and Serlachius tradition as a paper producer.

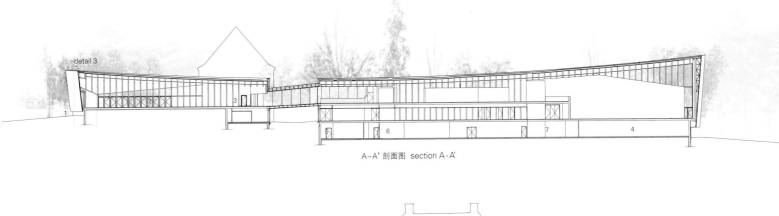

A-A' 剖面图 section A-A'

1.入口	1. entrance
2.集会大厅	2. assembly hall
3.入口门厅+ 交流区	3. entrance foyer+ communications
4.巡回展览区	4. travelling exhibition
5.冷冻区	5. freezer
6.小厨房	6. kitchenette
7.控制室	7. handling

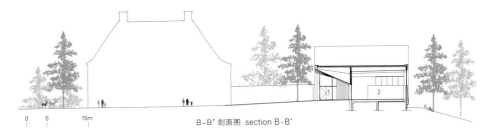

B-B' 剖面图 section B-B'

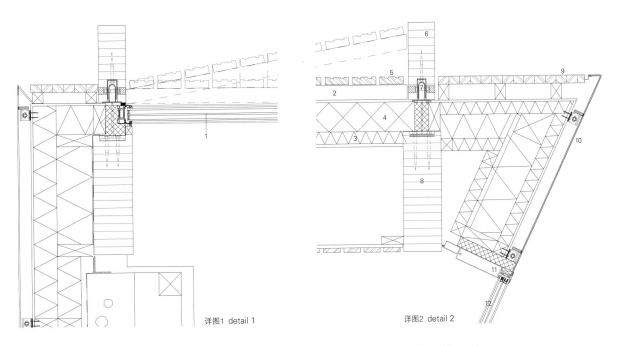

详图1 detail 1   详图2 detail 2

1. aluminium frame window with triple glazing 2. facade planks wooden horizontal support profile, painted in black color
3. thermal isolation with plasterboard 4. steel PU sandwich elements, painted in black 5. facade cladding – wooden planks
6. glulam wooden mullion 7. stainless steel mullion – column connectors 8. wooden glulam column 9. wooden board
10. glossy painted aluminium sheet as glass facade corner cladding 11. aluminium mullion 12. mullion-transom facade – triple glazing

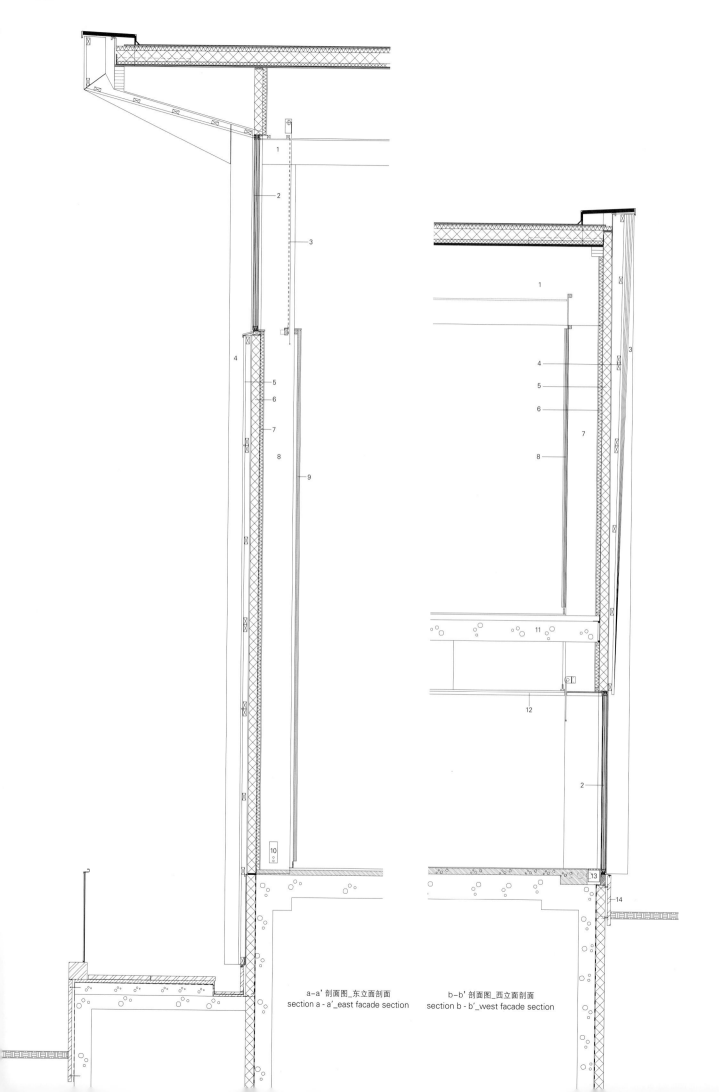

a-a' 剖面图_东立面剖面
section a - a'_east facade section

b-b' 剖面图_西立面剖面
section b - b'_west facade section

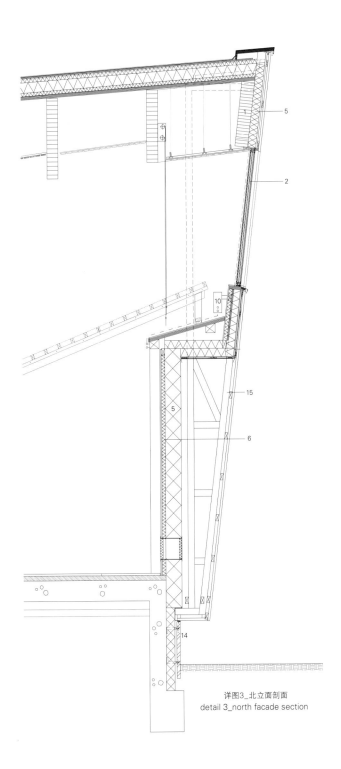

详图3_北立面剖面
detail 3_north facade section

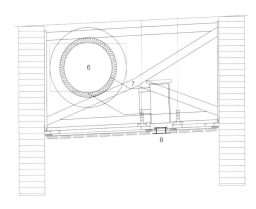

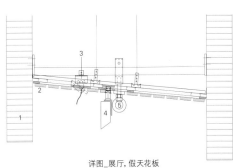

详图_展厅, 假天花板
detail_exhibition hall, false ceiling

1. glulam wooden beam
2. aluminium frame windows with triple glazing
3. sunscreen and removable wooden blackout panel
4. glulam wooden facade mullion
5. facade cladding – wooden planks
6. steel PU sandwich elements as thermal isolation and facade fire protection
7. thermal isolation between columns with double plasterboard
8. glulam wooden column
9. wooden LVL board covered with single plasterboard as "artwall"
10. radiator-in-wall and window heating
11. concrete slab carried by wooden columns
12. false ceiling from wooden profiles with acoustic absorbtion
13. trench heating with metal grill
14. grey granite stone plinth
15. prefabricated "V" shape lattice girder for facade cladding

1. glulam wooden beam
2. registrable panels cladded with wooden planks and grey acoustic absorption felt. wood is treated with fire retardant painting and semi transparent white painting
3. sockets for electricity and data with wooden cover
4. rail for light projectors
5. removable metal hook for hanging artworks
6. air duct
7. metal connector between wooden glulam beams
8. linear slot air diffuser

木材——诗意与实用 Wood – Poetic and Practical Possibilities

Knarvik社区教堂
Reiulf Ramstad Arkitekter

这座新Knarvik社区教堂位于风景优美的挪威卑尔根北部西海岸，建在可眺望文化遗址以及镇中心的生活便利的场地内。建筑静静地融入自然与人工环境之间的现有山坡中，为教堂提供了启发人灵感的石南属植物的景观环境。与众不同、极具创新的特点和地处中心的位置，使该建筑成为社区的地标性建筑。教堂面向所有人开放，人们可以在任何一天来到这里增强自己的信仰。教堂的功能通过其神圣的庄严性以及被大众认可的外形得以显示。通过高度不断上升的屋顶平面，教堂的尖顶、圣殿和小教堂都得以凸显。受挪威木条教堂的当地传统的启发，该建筑采用的是简洁明了的外形、材料和施工方法。紧凑的建筑体量在长方形的平面内分成两层，将这一神圣的空间从上到下分为文化区和行政区。

内部的教堂广场利用一个中庭楼梯将两个楼层连接起来，从而形成连续的空间。通过滑动玻璃墙，该空间还可以与圣殿合为一体或者分开，使空间可容纳500多人。木材是这一项目的主要材料，统一的外覆层使用的是预风化松木芯材，而所有内墙表面都采用浅色松木饰面。如同柳叶刀般又长又窄的窗户在能够允许日光进入室内的同时还能减少炫光。夜晚，从教堂内部散发的暖光向人们展示了其内部的宗教和文化活动。教堂意图为孩子们和年轻人提供一个安全的平台，成为他们聚会和培养信仰的场所，同时促进当地艺术、音乐和文化的发展。Knarvik社区教堂通过其自身的建筑风格、空间设计方案和材料使用，把宗教、文化以及当地特有的环境整合在一起。

Community Church Knarvik

The new Community Church in Knarvik, located on the scenic west-coast of Norway, north of Bergen, is built on a privileged site overlooking the cultural landscape and local town

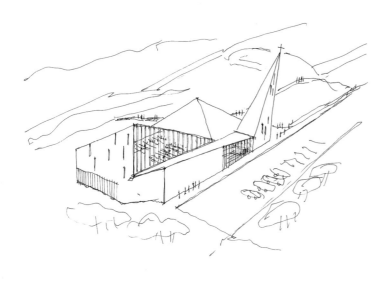

centre. The building is carefully adapted to an existing hillside between built and natural environment, providing the church with an inspiring context of the surrounding heath landscape. Its distinctive and innovative character and central location make itself a landmark in the community, to be inviting and inclusive for all people to cultivate their faith throughout the week. The church signals its function with a sacral dignity and recognisable form, where the church spire, sanctuary and chapel are emphasised by ascending roof planes. Inspired by the local tradition of Norwegian stave churches, the building utilizes clear geometries, materials and constructions. The compact building volume is split into two stories on a rectangular plan, separating the sacred spaces above from the cultural and administrative functions below. An internal "church square" connects the two levels with an atrium stair into a continuous space, and may be joined or separated from the sanctuary with sliding glass walls to accommodate more than 500 people. Wood is the key material of the project, expressed in the homogeneous cladding of pre-weathered pine heartwood and by the light-coloured pine finish on all interior surfaces. The building permits daylight into its volume through lancet-reminiscent tall and narrow windows, splayed in plan to reduce glare. At night, the warm glow of the interior reveals the activities of its religious and cultural events. The church aspires to provide a platform for a safe upbringing for children and youth, to become a local venue for gatherings and faith, and to facilitate art, music and cultural development. The Community Church Knarvik has an architectural expression, spatial solutions and materiality which unite religion, culture and the sitespecific context into a whole.

传统 tradition

Stave church
Wood as a building material
Church integrated the natural landscape
Church as an important cultural place
Tower with a spire

创新 innovation

7-day square - open to public / festivities every day
Diverse gathering spaces for all
Environmentally friendly and energy conscious
Platform for culture and art

Knarvik is rooted both in tradition
and forward-thinking solutions

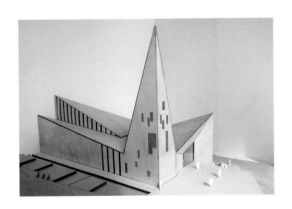

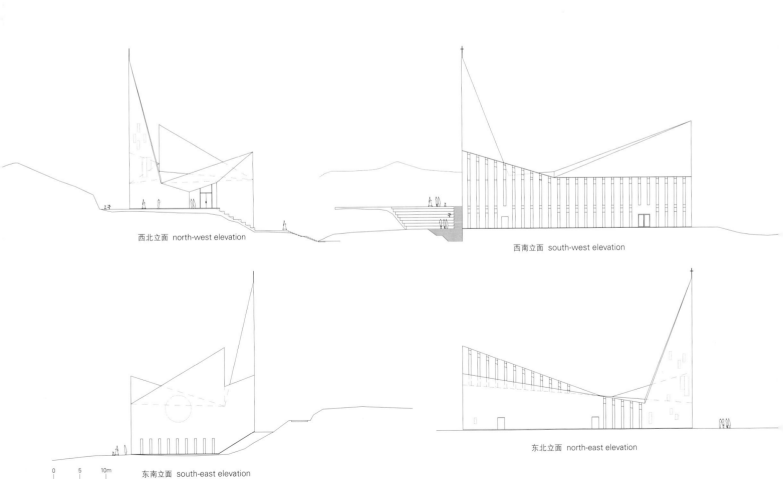

西北立面 north-west elevation

西南立面 south-west elevation

东南立面 south-east elevation

东北立面 north-east elevation

0 5 10m

项目名称：Community Church Knarvik
地点：Knarvik, Hordaland, Norway
建筑师：Reiulf Ramstad Arkitekter
甲方：Lindås Kyrkjelege Fellesråd
功能：new community church with cultural and administration facilities
用地面积：15,000m²
总建筑面积：1,100m²
有效楼层面积：2,250m²
造价：NOK 80M
委托类型：Invited competition(2010), 1st prize
设计时间：2010
施工时间：2014
竣工时间：2014
摄影师：©Hundven-Clements Photography (courtesy of the architect)

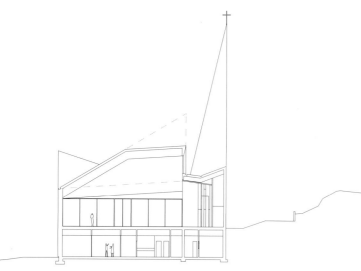

A-A' 剖面图 section A-A'

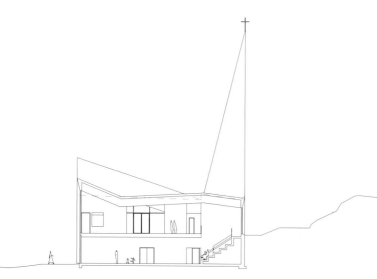

B-B' 剖面图 section B-B'

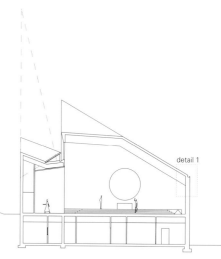

C-C' 剖面图 section C-C'

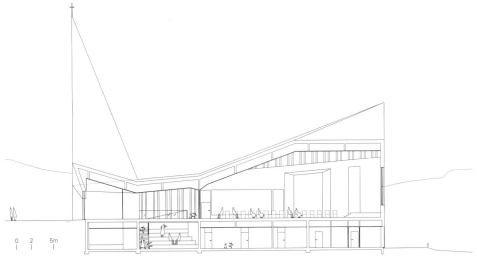

D-D'剖面图 section D-D'

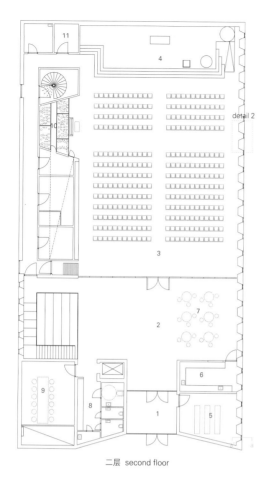

二层 second floor

1. 走廊
2. 门厅
3. 中殿
4. 圣坛
5. 教堂
6. 厨房
7. 咖啡室
8. 更衣间/卫生间
9. 会议室
10. 管风弦乐器室
11. 储藏室

1. vestibule
2. foyer
3. nave
4. altar
5. chapel
6. kitchen
7. cafe
8. wardrobe/wc
9. meeting room
10. organ
11. storage

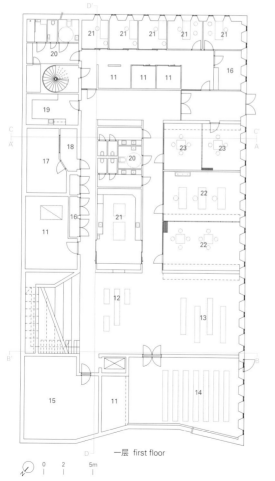

一层 first floor

12. 圆形剧场/舞台
13. 活动室
14. 礼堂
15. 技术间
16. 接待处/档案室
17. 录音室
18. 控制室
19. 洗衣房
20. 更衣间/卫生间/淋浴间
21. 办公室
22. 教室/会议室
23. 会议室/小组讨论室

12. amphitheater/stage
13. event space
14. auditorium room
15. technical room
16. reception/archives
17. recording
18. control room
19. laundry
20. wardrobe/wc/shower
21. office
22. teaching/meeting room
23. conference/group room

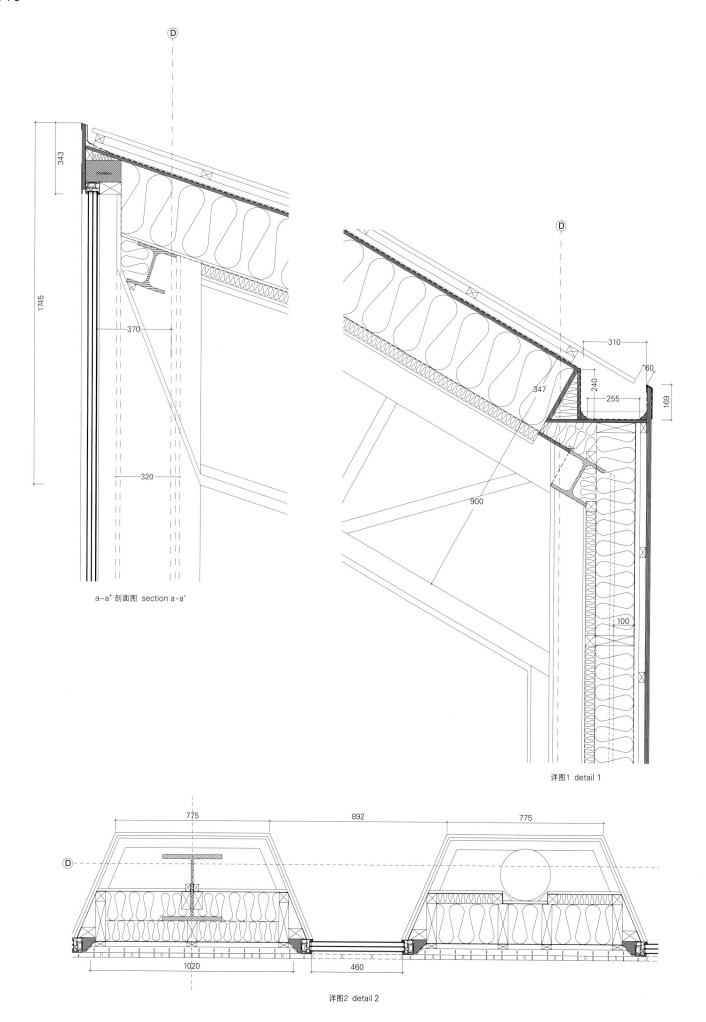

a-a' 剖面图 section a-a'

详图1 detail 1

详图2 detail 2

Créteil教堂的扩建项目
Architecture-Studio

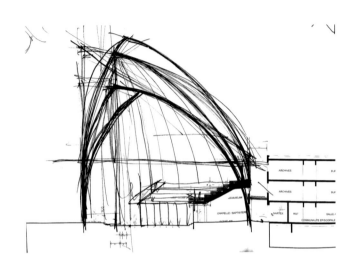

Créteil Cathedral Expansion

　　2009年，在Santier主教发起，Créteil教区协会选取了一个富有创意的方案，来为Créteil圣母院教堂进行扩建。该建筑由Charles-Gustave Stoskopf构思（他是罗马大奖赛的获得者），是一座典型的20世纪70年代建造的、"博采各家神学之长"的建筑。如今，它成为Créteil城现代遗产的一部分。其委托任务旨在将原有的使用面积扩大一倍，同时增强教堂在这座城市的辨识性。该设计与其说是对教堂进行翻新，不如说是进行重新开发，使其无论是从象征意义来说，还是教化方面来说，都被赋予了新的生命力。这座扩建后的教堂矗立在一个多文化交融的城市中，城内有五座天主教堂、十座犹太教堂、一座清真寺、一座新教教堂、四座福音派教堂、一座寺庙以及一座巴哈教堂。

　　该建筑兼具着两种风格，既有不同之处，又有相通的地方。指向天空的穹顶是参考了原教堂的轮廓而设计的。

　　颇具人性化功能的入口处轮廓与宏大的建筑结合，重点体现在教堂中殿从两个球形木覆层中延伸出来，使祭坛上方仿佛有一双正在祈祷的手的设计方面。新空间内可以举办更大型的集会。圣殿被重新修建，长凳也按照半圆形的形状布置。在阳光下，两个木质壳体之间安装的彩色玻璃将彩色的光芒映射在圣殿中。夜晚，室内被照亮，成为活跃的基督教社区的标志。钟楼位于教堂前广场的角落，独立于主建筑而存在。其狭长的轮廓成为教堂入口的标志，且楼体的连续性被老钟楼里的三座钟打断。钟楼再现了建筑的城市规模，并且在住宅楼林立的城市中成为该城市的标志。原有树木被移走后，教堂前庭的视野变得更加开阔。新广场建在其对面，不仅为当地的居民提供了便利，也丰富了教会的生活。

In 2009, on the initiative of Bishop Santier, the diocesan association of Créteil, opted for an ambitious project to expand the cathedral of Notre-Dame de Créteil. Conceived by Charles-Gustave Stoskopf, holder of the Prix de Rome, this architecture is typical of the 1970s when "the theology of blending-in" prevailed at the time. It is part of the contemporary heritage of the City of Créteil. The commission was to double the capacity of the cathedral and to enhance its visibility towards the city. More than a renovation, this project involved a major redevelopment of the cathedral, giving it a new architectural lease on life from a symbolic and pastoral point of view. The new cathedral is anchored in a multicultural city, which includes five Catholic churches, ten synagogues, a mosque, a Protestant church, four Evangelical churches, a Buddhist temple and a Bahai assembly.

A dialogue between two different architectural styles, yet consistent, is established. The dome pointing skywards is based on the footprint of the original cathedral.
The silhouette of the entrance, on a human scale, is now

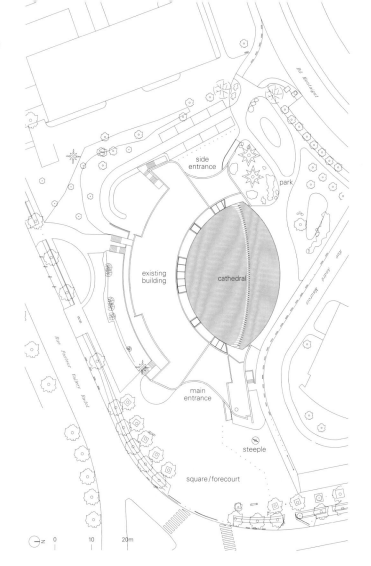

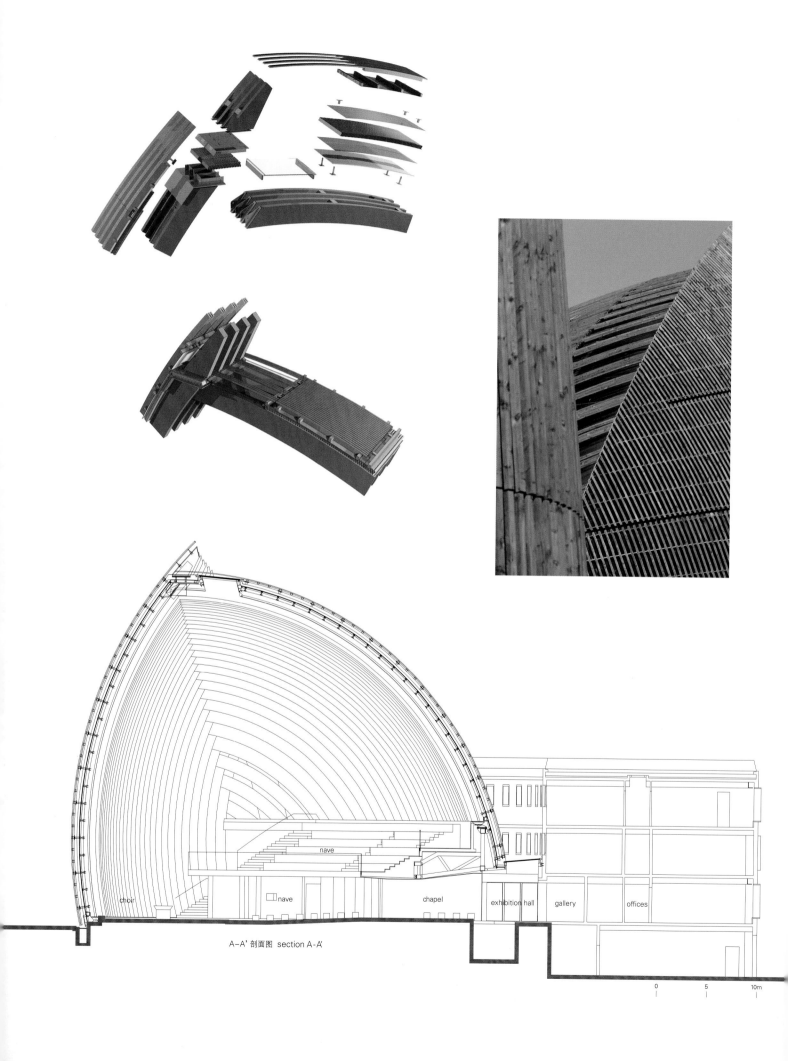

A-A' 剖面图 section A-A'

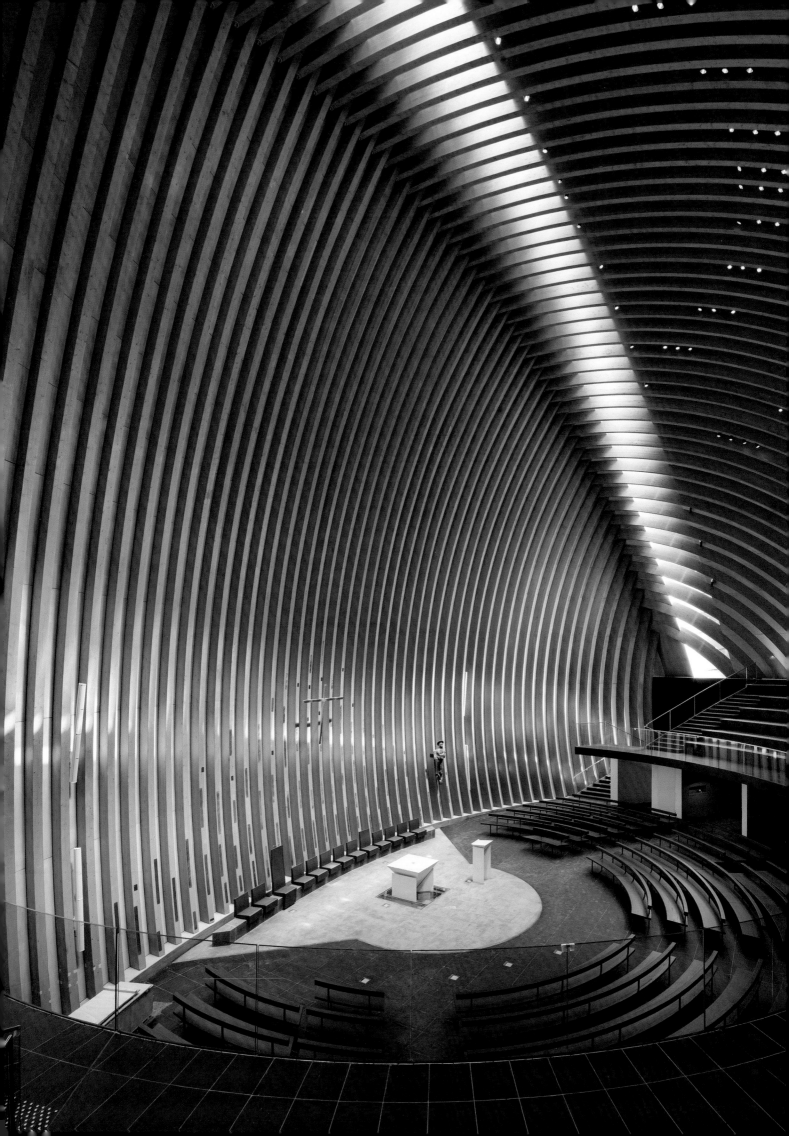

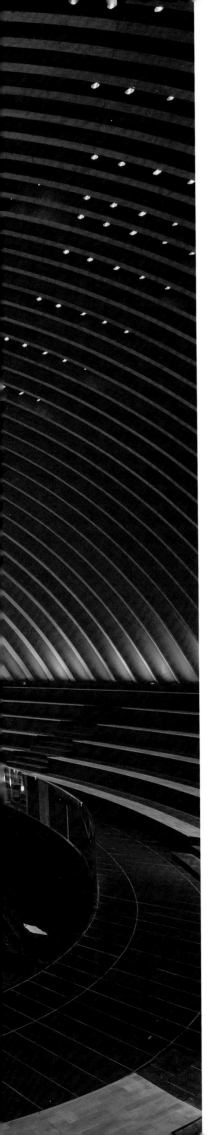

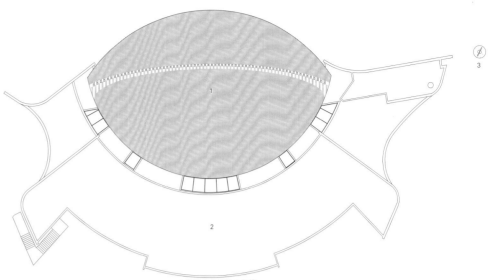

1.教堂 2.原建筑 3.钟楼
1. cathedral 2. existing building 3. steeple
屋顶 roof

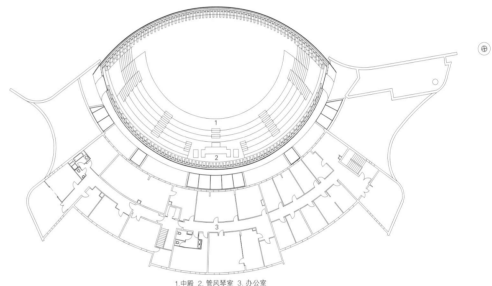

1.中殿 2.管风琴室 3.办公室
1. nave 2. organ 3. offices
二层 second floor

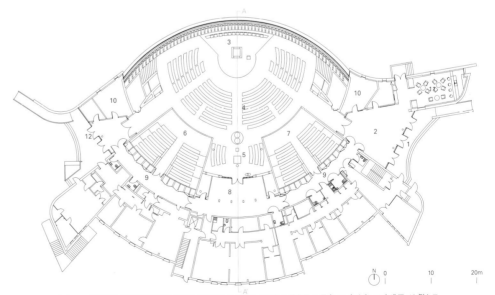

1.主入口 2.教堂前厅 3.唱诗班表演区 4.中殿 5.礼拜堂 6.会议室 7.礼堂 8.展览区 9.画廊 10.办公室 11.咖啡厅 12.侧入口
1. main entrance 2. narthex 3. choir 4. nave 5. chapel 6. conference room 7. auditorium
8. exhibition area 9. gallery 10. offices 11. cafe 12. side entrance
一层 first floor

joined with the monumental proportions of the new project, focusing on the nave of the cathedral that extends from two spherical wood-clad hulls, like two hands joined in prayer that meet above the altar. Large gatherings can be held in this new space. The existing sanctuary has been remodeled and the benches are placed in a semicircle. In daylight, the stained-glass window located at the junction of the two hulls shed a colored light onto the sanctuary, while at night, illuminated from inside, they become the symbol of a living Christian community. The steeple, detached from the building on the corner of the forecourt, marks the cathedral entrance with its slender silhouette, punctuated by three bells from the old campanile. It restores the building's urban scale and becomes a sign in the city beside the large residential buildings of the neighborhood. The view onto the cathedral forecourt is freed by opening the curtain of trees. The new square, built on the opposite side, is an amenity for local residents, and an extension of parish life.

项目名称：The expansion of the Cathedral of Créteil
地点：2 Rue Pasteur Vallery Radot, 94000 Créteil, France
建筑师：Architecture-Studio
承包商：Association diocésaine de Créteil
结构和覆层设计：T/E/S/S / 流体学设计：Louis Choulet / 音效设计：AVA
照明设计师：8'18" / 估算师：Éco-Cités / 木材专家：Sylva Conseil
参与设计人员：Leon Grosse, Fargeot Lamelle Colle (Arbonis), Cabrol
总建筑面积：1,400m² / 教堂高度：22.4m
表壳表面积：1,600m² / 金属框架质量：40t
总造价：9 million euros before tax
施工造价：5,75 million euros before tax
竞赛时间：2009 / 竣工时间：2015
摄影师：©Luc Boegly (courtesy of the architect) - p.119, p.120~121, p.122
©Yves Mernier (courtesy of the architect) - p.114, p.116, p.117, p.118, p.123, p.124~125

苏黎世动物园大象馆

Markus Schietsch Architekten

作为苏黎世动物园扩建项目总体规划的第一步,新建的康卡沾大象公园采用的规模和结构使其成为一座现代化的、贴近自然的建筑,可容纳十头大象。

根据现代动物园的理念,游客进入动物栖息地主要是为了进行一次全面的大自然体验。因此,建筑被构思为建筑和景观场景之间的相互作用,同时建筑和景观之间要建立共生关系。

所有可见的自然元素都尽可能地贴近自然。这种有机的、自由风格的建筑构造与自然植被相互交织,营造室内的氛围。新象馆及其室外空间主要位于岩壁脚下的一处植被茂密的景观中。其富有特色的元素是与景观交融的、引人注目的木屋顶,木屋顶采用的是扁壳结构,且外形不规则。屋顶嵌入了透明的迷宫式结构,与周围的丛林建立了一种有机的关系。在室内,屋顶展现了其大气效应:好像阳光透过树冠,穿过复杂的屋顶结构,产生了不断变化的光线环境。

利用建筑中心内部的组合构造,屋顶横跨了整个室内景观,组合构造包括密集植物装饰的景观路线、能观看大象游泳和潜泳的水下观赏区、游客住宿区以及培育小象的不可见的管理区。

屋顶被设计成薄薄的木质壳体,跨度约为80m,预制的交叉层压木板弯曲成型,然后用钉子钉在一起。洞口也是现场从巨大的木质壳体中切割形成的。

不断变化的立面结构包括顶部看起来与屋顶相接的木薄板,薄板如同一条有机带,展现了承重区域的位置。屋顶的壳体所呈现的图像与动态的立面形成了独具氛围的外围护结构和使人身临其境的自然构造,将设计的本质集中在建筑和景观之间的一种共生关系中。

Elephant House Zoo Zürich

As a first building block in the masterplan of the extension of the Zürich Zoo the newly designed Kaeng Krachan Elephant Park allows by its size and structure for a contemporary and close-to-nature approach in keeping the 10 elephants.
In accordance to modern zoo philosophy the visitor immerges into the habitat of the animals for a holistic nature experience. Therefore the architectural elements were conceived as an interplay of the landscape image with architecture and landscape entering into a symbiotic relationship.
All visible building components were designed as analogies of nature. The organic and free-formed structures interweave with the natural vegetation and induce the atmospheric interior space. The new elephant house with its exterior compounds is embedded in an extensive and densely vegetated landscape at the foot of a rock wall. The characteristic element of the new elephant house is its striking wooden roof which blends into the landscape as a shallow free-form shell-structure. The roof dissolves into a transparent maze-like

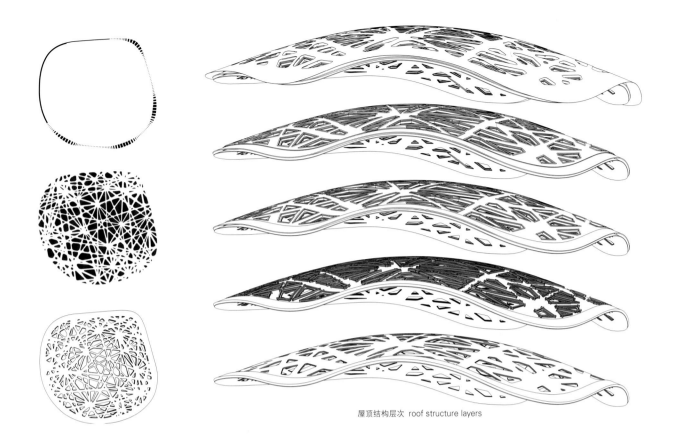

屋顶结构层次 roof structure layers

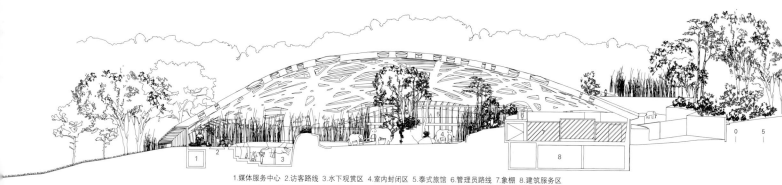

1.媒体服务中心 2.访客路线 3.水下观赏区 4.室内封闭区 5.泰式旅馆 6.管理员路线 7.象棚 8.建筑服务区
1. media services 2. visitors path 3. underwater inside view 4. inside enclosure
5. Thai lodge 6. keepers path 7. stables 8. building services
A-A' 剖面图 section A-A'

维修层_抬高的Kerto板
maintenance level_elevated Kerto plate

ETFE衬层, 蓄水层
ETFE cushions, aquifer

边缘肋板 edge ribs

通风空间 ventilation space

支撑外壳 supporting shell

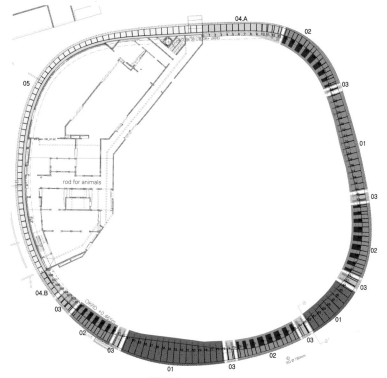

结构详图 structure detail

边缘固定的幕墙竖框区
clamped curtain wall mullions area

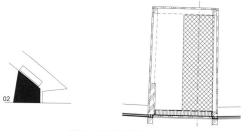

嵌板(屋顶荷载转移至此)
paneling area (transfer of roof loads)

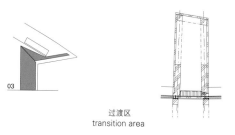

过渡区
transition area

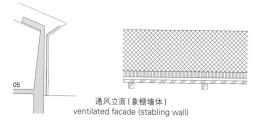

边缘固定的幕墙竖框区(象棚墙体)
clamped curtain wall mullions area
(stabling wall)

通风立面(象棚墙体)
ventilated facade (stabling wall)

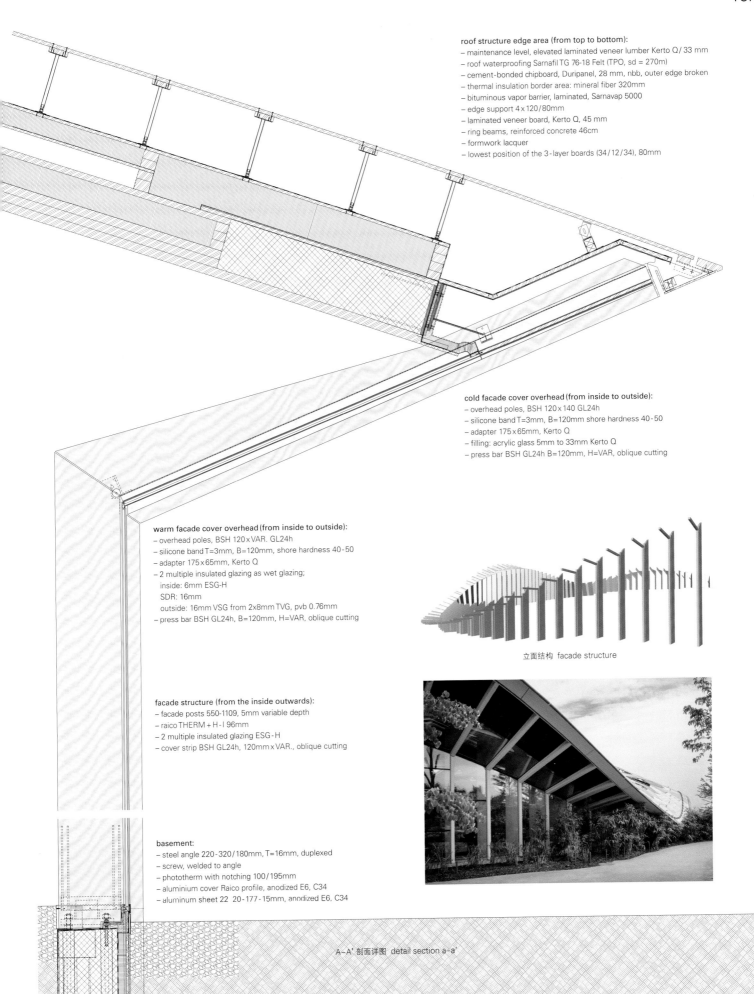

roof structure edge area (from top to bottom):
– maintenance level, elevated laminated veneer lumber Kerto Q/ 33 mm
– roof waterproofing Sarnafil TG 76-18 Felt (TPO, sd = 270m)
– cement-bonded chipboard, Duripanel, 28 mm, nbb, outer edge broken
– thermal insulation border area: mineral fiber 320mm
– bituminous vapor barrier, laminated, Sarnavap 5000
– edge support 4 x 120/80mm
– laminated veneer board, Kerto Q, 45 mm
– ring beams, reinforced concrete 46cm
– formwork lacquer
– lowest position of the 3-layer boards (34/12/34), 80mm

cold facade cover overhead (from inside to outside):
– overhead poles, BSH 120 x 140 GL24h
– silicone band T=3mm, B=120mm shore hardness 40-50
– adapter 175 x 65mm, Kerto Q
– filling: acrylic glass 5mm to 33mm Kerto Q
– press bar BSH GL24h B=120mm, H=VAR, oblique cutting

warm facade cover overhead (from inside to outside):
– overhead poles, BSH 120 x VAR. GL24h
– silicone band T=3mm, B=120mm, shore hardness 40-50
– adapter 175 x 65mm, Kerto Q
– 2 multiple insulated glazing as wet glazing;
  inside: 6mm ESG-H
  SDR: 16mm
  outside: 16mm VSG from 2x8mm TVG, pvb 0.76mm
– press bar BSH GL24h, B=120mm, H=VAR, oblique cutting

立面结构 facade structure

facade structure (from the inside outwards):
– facade posts 550-1109, 5mm variable depth
– raico THERM + H-I 96mm
– 2 multiple insulated glazing ESG-H
– cover strip BSH GL24h, 120mm x VAR., oblique cutting

basement:
– steel angle 220-320/180mm, T=16mm, duplexed
– screw, welded to angle
– phototherm with notching 100/195mm
– aluminium cover Raico profile, anodized E6, C34
– aluminum sheet 22 20-177-15mm, anodized E6, C34

A-A' 剖面详图 detail section a-a'

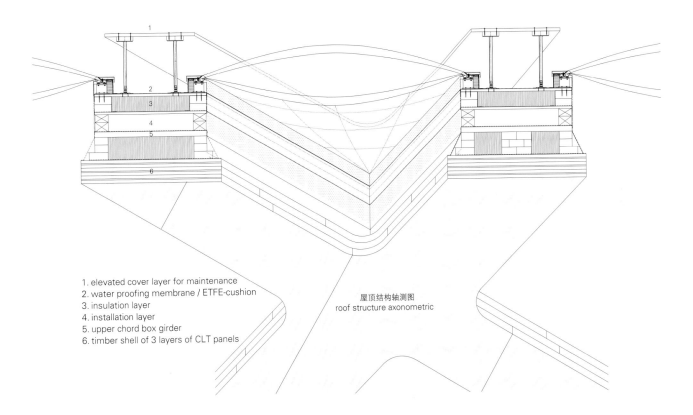

1. elevated cover layer for maintenance
2. water proofing membrane / ETFE-cushion
3. insulation layer
4. installation layer
5. upper chord box girder
6. timber shell of 3 layers of CLT panels

屋顶结构轴测图
roof structure axonometric

项目名称：Elephant House Zoo Zürich / 地点：Zürich, KT Zürich, Switzerland / 建筑师：Markus Schietsch Architekten GmbH / 项目管理：cga gmbH, Winterthur, BGS Architekten / 结构工程师：Walt + Galmarini AG / 建筑服务工程师：TriAir Consulting AG / 电气工程师：Schmidiger + Rosasco AG / 景观建筑师：Lorenz Eugster Landschaftsarchitektur GmbH / 景观施工管理：Vetsch Partner Landschaftsarchitekten AG / 技术服务：TriAir Consulting AG / 施工管理：Fischer Architekten AG, BGS Architekten / 照明设计：Bartenbach Lichtlabor AG / 甲方：Zoo Zürich AG / 有效楼层面积：8,440 m² / 体积：68,000m³ / 造价：57.0 Mio SFr / 施工时间：2008—2014 / 摄影师：©Dominique Marc Wehrli (courtesy of the architect)-p.126~127, p.129, p.130, p.137, p.139, ©Andreas Buschmann-p.131, p.132~133, p.134, p.135, p.138, p.140~141

structure that establishes an organic relationship to the surrounding forest. In the interior the roof unfolds its atmospheric effect: as if through a canopy of trees the sunlight filters through the intricate roof structure generating constantly changing light atmospheres.

The roof spans the interior landscape with the central inner compound that is framed by the densely vegetated visitors path, as well as an underwater viewpoint to see the gentle giants dive and swim, a visitors lodge and the management area for the fostering of the elephants that is not visible to the visitor.

The roof is designed as a shallow wooden shell with a span of

80 meters. Prefabricated cross laminated timber panels were bent on-site into form and nailed up. The openings were cut out on-site from the massive wooden shell.
The continuously changing facade structure consists of lamellas that seemingly grow up to the edge of the roof as an organically shaped band indicating the loadbearing areas.

The iconographic shell of the roof together with the dynamic facade form an atmospheric envelope and pictographic "Nature-Construction" concentrating the essence of the design into a symbiosis between architecture and landscape.

Sognefjellshytta高山酒店的新入口
Jensen & Skodvin Architects

木——诗意与实用 Wood – Poetic and Practical Possibilities

该场地位于挪威最高山路的最高点,即低矮的内陆和沿海地区之间,这条山路自古就有,并且一直不断地开发,直到1938年左右才形成了目前的形态。该地区吸引了众多的游客,而Sognefjellshytta小旅馆自20世纪30年代开始就已经存在了。随着游客的增加,房间的数量也逐渐增加。2009年,我们被邀请去参观入口处的设施。我们提议通过使用一个大型预制木质连接结构来将两座现存的建筑联系在一起。由于夏季非常短暂(大约3个月,之后道路由于雪的问题封闭),因此这个项目的施工时间足足经历了四个夏季。预制结构的做法是十分明智的。我们在电脑上通过3D绘图使用电子锯来完成这一结构的设计。然后用一个夏天来安装结构,关闭建筑,剩下的两个夏季进行内部装修,且不耽误酒店运营。

Sognefjellshytta High Mountain Hotel's New Entrance

The site is located at the highest point on the highest mountain pass road in Norway, between the lower inland and the coastal areas. The mountain pass path/road is ancient, and has been developed gradually until it reached its current form in the around 1938. It has become a huge tourist attraction and the small hotel Sognefjellshytta has been existing since the

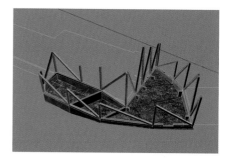

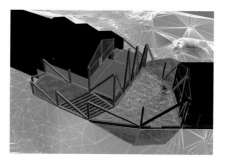
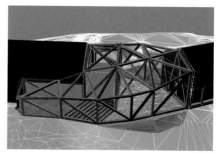

项目名称：Tourist project Sognefjellshytta high mountain hotel / 地点：Sognefjellet, Norway / 建筑师：Jensen & Skodvin Architects
项目团队：Jan Olav Jensen_project leader, Torunn Golberg_project leader, Øystein Skorstad, Thomas Knigge / 景观/室内建筑师：Jensen & Skodvin Architects
静态设计顾问：Siv. Ing. Finn Erik Nilsen / 甲方：Sognefjellshytta / 竣工时间：2015 / 摄影师：courtesy of the architect

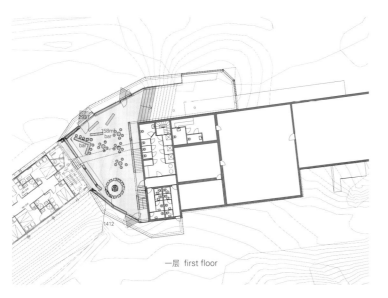
一层 first floor

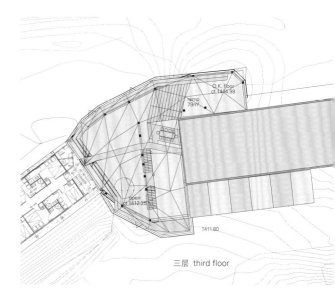
三层 third floor

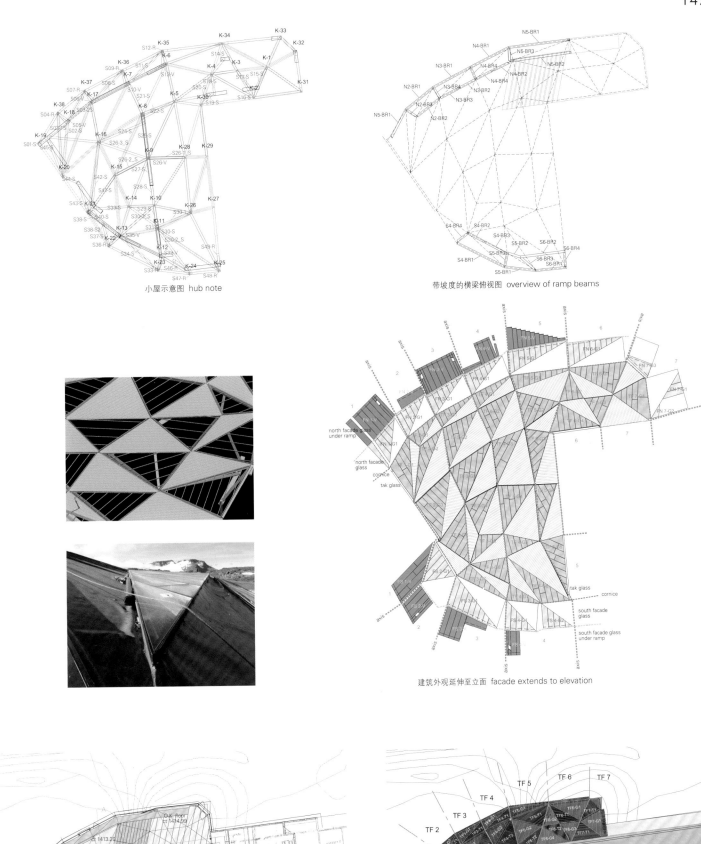

小屋示意图 hub note

带坡度的横梁俯视图 overview of ramp beams

建筑外观延伸至立面 facade extends to elevation

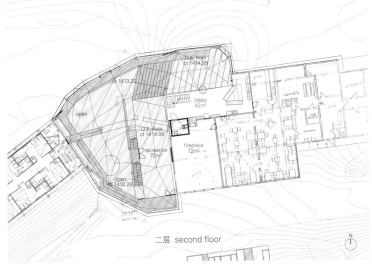

二层 second floor

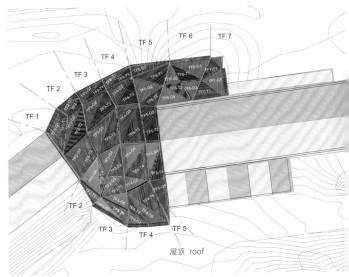

屋顶 roof

thirties. It has gradually been developed with more rooms as the number of guest has increased. In 2009 we were asked to look at the entrance facilities. We proposed to connect the two existing buildings with a large connecting prefabricated structure in wood. As the summer season is very short (around 3 months, then the road is closed due to too much snow), four summers have been used to build the project. It was obviously rational to prefabricate the structure. This was done directly from our 3D drawings with a computerized saw. One summer was used to build the foundations, one summer to erect the structure and close the building, then the next two summers to do the interior, while the hotel was running.

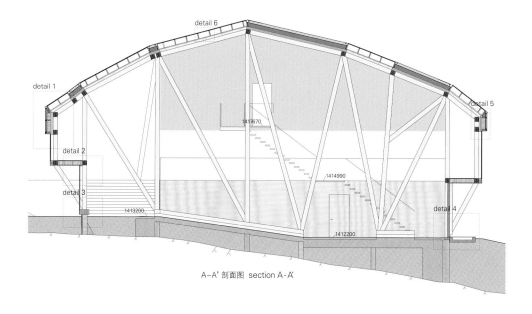

A-A'剖面图 section A-A'

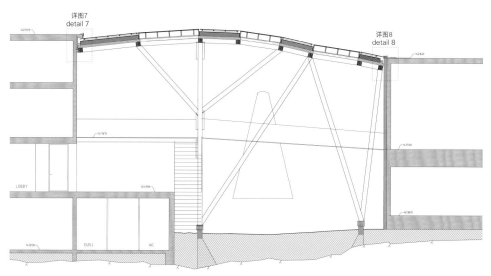

B-B'剖面图 section B-B'

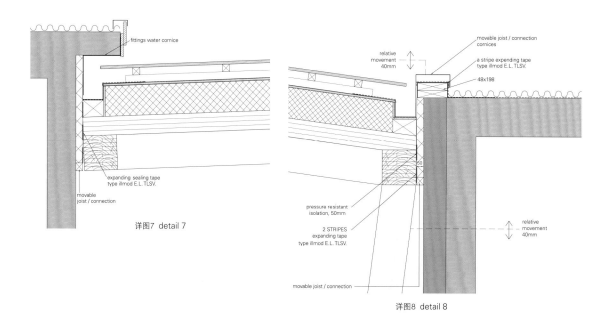

详图7 detail 7

详图8 detail 8

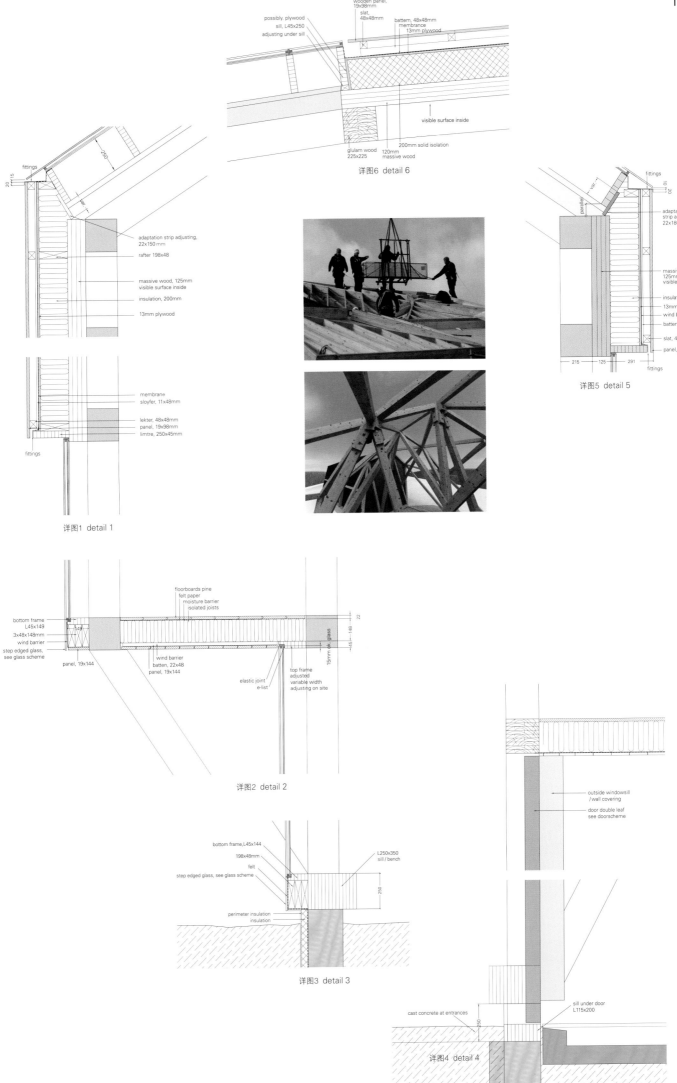

扎哈·哈迪德：
光辉人物的精神遗产
Zaha Hadid: The Legacy of a Star

"作为一名建筑师，她一直在拓展建筑和城市设计的范围。她的作品对新的空间概念加以试验，突出了现有的城市景观的存在感，她的作品涵盖了设计的各个领域，从城市规模设计到室内，再到家具设计。"1

全球性的冲击

扎哈·哈迪德于2016年3月31号去世的消息给全球的建筑协会带来了冲击。世界各地颇具盛名的建筑专业人员以及建筑机构都表达了他们对扎哈·哈迪德的尊敬，同时对她的作品给予了盛赞。他们的言语中透露出了对扎哈·哈迪德作品的钦佩，无论是她在世界各地所设计的作品，还是她遗留的作品。

她的同事——扎哈·哈迪德事务所的主管帕特里克·舒马赫写到："对于扎哈的去世，我们都感到十分震惊和惋惜，她是一位美丽、独立、有天分的领导和朋友。"2

"Known as an architect who consistently pushes the boundaries of architecture and urban design, her work experiments with new spatial concepts intensifying existing urban landscapes and encompassing all fields of design, from the urban scale to interiors and furniture".1

A Global Shock

The news of Zaha Hadid's sudden death on 31st of March left the global architecture community in shock. Many of the most remarkable figures of the profession and architecture institutions from around the world promptly showed their respect and paid tribute to her figure. Their words portrayed great admiration for what her figure has represented within the world of architecture and the unique legacy that she now leaves behind.

Her colleague and Director at Zaha Hadid Architects, Patrick Schumacher wrote *"We are all shocked and devastated that we lost Zaha today, a most beautiful, individual, talent leader and friend."*2

Norman Foster said: *"I am devastated by the news of the loss of Zaha Hadid and cannot comprehend the enormity of her passing away. I became very close to her as a friend and colleague in parallel with my deep respect for her as an architect of immense*

诺曼·福斯特说道："我十分震惊于扎哈的去世，不能从她的过世中释怀。我既是她亲密的朋友，又是共同工作的同事，我对于她十分尊重，她是一位伟大的建筑师，她的存在对于全球的建筑界都极具意义。"3 理查德·罗杰斯说道："她是一位伟大的建筑师，一名杰出的女性，一位杰出的人物。在过去几十年新兴的建筑师里，她比任何一个人产生的影响都要大。作为一名女性，她在自己的道路上不断奋斗。她也是普利策奖的第一位女性获得者。"4 丹尼尔·利贝斯金德说："我深深惋惜一位伟大的建筑师，我的好同事在今天去世了。扎哈的精神将永远留在她的作品及工作室中，我们的心永远在一起。"5 弗兰克·盖里是这样描述扎哈的："她是一位非常伟大的建筑师，也是一位好的朋友，我将永远怀念她。"6 史蒂文·霍尔说："她的建筑精神与能量将会与她所设计的那些不可思议的建筑同在。我们深深地怀念她。"7 比雅可·因戈尔说："听到扎哈去世的消息，我们都感到十分悲痛。我们的想法与她家人、同事、朋友的想法是一致的。愿她安息吧！"8 隈研吾写到："扎哈·哈迪德是一位伟大的建筑师，后现代主义时期以后，扎哈引领了建筑的潮流……超越她的成就是留给下一代建筑师的挑战，包括我自己在内。"9

巴格达、贝鲁特、伦敦和世界

为了更好地理解她的个人形象，以及她的职业形象是如何随着岁月而发展的，我们需要好好了解一下她的背景。1950年，她出生在巴格达一个受过良好教育的伊斯兰家庭，这个家庭深受西方多元主义文化的影响。就如事实所反映的那样，她在一个天主教学校上学，但该学校实际上实行的却是宗教多元化。后来，她搬到了黎巴嫩首都贝鲁特，在美国大学学习数学。1972年，22岁的扎哈·哈迪德搬到了英国，在伦敦她参加了那里的建筑协会。当时伦敦的建筑文化环境是比较活跃的，像雷姆·库哈斯、埃利亚·增西利斯和伯纳德·屈米等建筑师都积极地参与教学和公共辩论。在建筑协会，扎哈·哈迪德遇到了OMA未

stature and global significance."3 Richard Rogers described her as: *"A great architect, a wonderful woman and wonderful person…Among architects emerging in the last few decades, no one had any more impact than she did. She fought her way through as a woman. She was the first woman to win the Pritzker prize."*4 Daniel Libeskind said himself to be *"devastated by the loss of a great architect & colleague today. Her spirit will live on in her work and studio. Our hearts go out."*5 Frank Gehry wrote about Zaha Hadid: *"She was a great architect and a great friend. I will miss her."*6 Steven Holl described her as *"an amazingly original spirit in architecture and her energy will live on through her incredible buildings. We will miss her deeply."*7 Bjarke Ingels held: *"we are all incredibly sad to hear about the loss of Zaha Hadid. our thoughts are with her family, friends and colleagues. May she rest in peace".*8 Kengo Kuma published: *"Dame Zaha Hadid was a great architect who led the world of contemporary architecture after the era of Postmodernism […]To surpass her achievement is the challenge left to the next generation of architects following her – myself included."*9

Baghdad, Beirut, London and The world

In order to better understand her particular figure and how her professional profile evolved along the years, it is important to look at her background. Born in Bagdad in 1950 to a well-educated Islamic family oriented toward Western multiculturalism, as shown in the fact that she attended a Catholic school but which was religiously diverse, she moved to the Lebanese capital Beirut to study maths at the American University. In 1972, aged 22, Zaha Hadid moved to the United Kingdom, where she enrolled in the Architectural Association in London. The architectural cultural environment in London at that time was particularly stimulating, with architects like Rem Koolhaas, Elia Zenghelis and Bernard Tschumi involved in teaching and public debates. At the Architectural Association Zaha Hadid liaised with the future founding partners of OMA, which she joined after completing her studies. These acquaintances were significant for the development of Zaha early career. Koolhaas described her as a *"combination of beauty and strength, […] incredibly generous and incredibly funny, […] She was basically family"*10. Zaha Hadid has always combined work in the practice with academic teaching. In the 1990s she held the Kenzo Tange Chair at Harvard University, the Sullivan Chair at the School of Architecture of the University of Illinois, and a postgraduate studio at Columbia University. In 2001 Zaha Hadid started teaching at the University of Applied Arts Vienna in Austria. After working at OMA she opened her

来的合伙创始人，在完成学业后，扎哈便把自身投入到了OMA中。这些熟人对于扎哈职业生涯的早期发展十分重要。库哈斯把她描述为一个"力与美的组合……有着令人难以置信的慷慨和风趣……就像家人一样"。[10] 扎哈总是将实践和学术教学相结合。在20世纪90年代，扎哈先后在哈佛执掌过丹下健三教席，在伊利诺斯大学设计学院执掌沙利文教席，在哥伦比亚大学的工作室工作。2001年，扎哈开始在奥地利的维也纳应用艺术大学执教。1980年，在离开OMA事务所后，她在伦敦开始了自己的建筑实践。从那时起，随着世界各地分公司的开设，办公室的数量成倍增长。到目前为止，她在44个地区、55个国家先后建造了950个项目。她的员工人数为400人左右。[11]

在她长达36年的职业生涯中，扎哈·哈迪德获得了国际上对她杰出建筑成就的认同。她获得了几个重要的奖项和奖金，包括2004年的普利策建筑奖、2010年和2011年的斯特灵奖，以及2016年3月的英国皇家建筑师协会金奖。

绘图

在成为国际赫赫有名的建筑师以前，扎哈因其颇具创新力的绘图在建筑领域闻名。在那段时间，她被人们认为是"纸上谈兵"。[12] 这些图纸中有许多是设计大赛的参赛作品。支离破碎、富有活力的建筑碎片以及景观元素充斥在整个画布上，被无形的力量聚集在一起。极具形状的彩色平面被分离，散开，然后又重新聚集在一起。这些作品在描绘拟定的建筑方案时用了一种非常新颖、激进的方式，无法与其他建筑师的作品相比较，也与传统方法有明显的区别。在她的绘图中，建筑方案与城市环境通过使用新型规划系统展现出来。那些违反万有引力的结构平面被放置在一条可无限延长的水平线上，这条水平线可以将所有的元素都准确地容纳其中。这些早期的绘图为扎哈·哈迪德为项目设想的空间定义提供了模式，不过在针对设想的空间的精确性方面，这些作品也留有了一定的余地。1983年，她针对香港山顶休闲俱乐部所提出的规划被誉为"地质至上主义"[13]。她提议建造一座"人工抛光

own architecture practice in 1980 in London. Since then the office has grown exponentially with branch offices all over the world. To date, her work included 950 projects in 44 countries and 55 nations, and her practice counts around 400 staff.[11] Over the course of her 36-year long career Zaha Hadid gained international recognition for her outstanding achievements in architecture. She received several important awards and prices, including the Pritzker Architecture Prize in 2004, the Stirling Prize in 2010 and 2011, and the RIBA Royal Gold Medal in March 2016.

Drawings

Before becoming an international star-architect, Zaha Hadid was known in a few architectural niches chiefly for her innovative drawings. At that time, she was *"seen as something of a paper tiger"*[12]. Many of these drawings were design competition entries. Fragmented and animated building pieces and landscape elements float over the whole canvas, flying and coming together attracted by invisible forces. Shaped coloured planes were broken up, spread out and reassembled all at once. They depicted the proposed architectural intervention in a radical new way, hardly comparable with other architects' drawings and clearly different from traditional methods.

In her drawings the building proposal and the urban context were represented by using new systems of projection: constructive planes defying gravity were dictated by a powerful and infinite horizon line that put together the elements in perspective. These early drawings formulated the spatial terms that Zaha Hadid envisioned for the project proposal, yet they allowed certain room for interpretation regarding what the definite and precise proposed space might be. The proposal she made in 1983 for The Peak Leisure club in Hong Kong, is described as a *"suprematist geology"*[13]. She proposed a "man made polished granite mountain", a city landmark that cut through one of the mountains surrounding Hong Kong like a knife. Her winning proposal for The Peak was never built but the competition material has been widely published on architecture media, and put the spotlight of the international architecture scene on Zaha Hadid.

Her drawings have been so successful and interesting that she was invited to participate in the Deconstructivist Architecture exhibition at the Museum of Modern art (MoMA) in New York in 1988, organized by Philip Johnson and Mark Wigley. The exhibition also featured works by Frank Gehry, Daniel Libeskind, Rem Koolhaas, Peter Eisenman, Coop Himmelb(l)au, and Bernard Tschumi. The participation in the Deconstructivist Ar-

花岗岩山"作为一个城市地标，这座地标像一把刀一样穿过环绕香港的某一座山。然而，这一获奖方案到最后并未真正实施，但是比赛所用的材料却被建筑传媒广泛报导，也使扎哈·哈迪德成为了国际建筑行业的焦点。

她的绘图十分成功、有趣，也因此在1988年，她被邀请参与了由菲利普·约翰逊和马克·威格利组织的在纽约现代艺术博物馆（MoMA）举办的解构主义建筑展（这个展览还展出了弗兰克·盖里、丹尼尔·利贝斯金德、雷姆·库哈斯、彼得·艾森曼、库柏·西梅布芬事务所和伯纳德·屈米的作品）。参与解构主义建筑展是扎哈·哈迪德职业生涯的一个里程碑。虽然在那个时候，她还没有一个独立完成的项目，但她依然是享誉全球的建筑师。

正如在现代艺术博物馆展览的新闻稿中写的那样：解构主义建筑展将重点放在七位国际建筑师身上，这些建筑师最近的作品代表了建筑方面一个新的灵感的出现。这些建筑师意识到了现代社会的不完美，并且在寻求解决方案。用约翰逊的话来说，就是享受"拘束感所带来的乐趣"，他们故意违反现代主义的立方体和直角，痴迷于对角线、圆弧和扭曲的弧线……绘图中不再带有传统的和谐、统一和清晰感，取而代之的是不和谐、破碎和神秘感。14

从纸张到混凝土

许多年来，她的绘图在全球范围内的建筑杂志上发表，但这些作品都被认为是单纯的纸上设计，没有成为混凝土实体建筑：不切实际，太激进，不适合操作。然而，在20世纪90年代初期，这一见解发生了戏剧性的变化。1993年在德国莱茵河畔威尔城建造的维特拉消防站是扎哈职业生涯的转折点。这是第一次她将绘图中浮动的彩色平面通过混凝土和有形的空间，打造成一座实体建筑。

1998年，当扎哈·哈迪德被委任设计位于俄亥俄州辛辛那提的理查德&洛伊丝·罗森塔尔当代艺术中心时，她终于获得了国际的认可。

chitecture set a milestone in Zaha Hadid's professional career. Although, at that time, she had no single project built yet, her name was amongst well-established global architects. As explained in the MoMA press release of the exhibition: Deconstructivist Architecture focuses on seven international architects whose recent work marks the emergence of a new sensibility in architecture. The architects recognize the imperfectability of the modern world and seek to address, in Johnson's words, the *"pleasures of unease"*. Obsessed with *diagonals, arcs, and warped planes, they intentionally violate the cubes and right angles of modernism. [...] The traditional virtues of harmony, unity, and clarity are displaced by disharmony, fracturing, and mystery.*[14]

From paper to concrete

For many years her drawings were published in architecture magazines globally, yet they were considered chiefly as mere bi-dimensional speculations, far from having a concrete architectural dimension: unrealistic, too radical and impossible to build. However, in the early 1990s this opinion was about to change dramatically. The construction of the Vitra Fire Station in Weil am Rhein, Germany, in 1993 represented a turning point in Zaha Hadid's career. For the first time the floating coloured planes of her drawings acquired a concrete and tangible space, becoming a physical building.

Zaha Hadid achieved international recognition in 1998 when she was commissioned to design the Lois & Richard Rosenthal Center for Contemporary Art in Ohio, Cincinnati. The 8500m^2 art center, which opened to the public in 2003, was conceived as a dynamic public space. It was described as an *"urban carpet drawing in pedestrian movement from surrounding areas, running from the building's exterior through the entrance, lobby and on into the interior."*[15] She described the spatial experience of the Center as *"a new fluid kind of spatiality of multiple perspective points and fragmented geometry, designed to embody the chaotic fluidity of modern life."*[16]

On the three above mentioned projects, namely The Peak in Hong Kong, the Vitra Fire Station in Germany and the Contemporary Art Center in Ohio, Zaha Hadid pursued a fluid development of the spaces that related the architecture with its urban context and landscape. In these early projects Zaha Hadid formulated her interest in the interface between architecture, landscape and topography. This relationship of forms and context would be one of the main characteristics of her work over the course of the following years.

In the approximately fifteen years since the Lois and Richard

维特拉消防站，莱茵河畔威尔城，德国，1993
Vitra Fire Station, Weil am Rhein, Germany, 1993

Phaeno科学中心，沃尔斯堡，德国，2005
Phaeno Science Center, Wolfsburg, Germany, 2005

宝马中心工厂，莱比锡，德国，2005
BMW Central Building, Leipzig, Germany, 2005

MAXXI 21世纪艺术博物馆，罗马，意大利，2009
MAXXI Museum of XXI Century Arts, Rome, Italy, 2009

广州大剧院，广州，中国，2010
Guangzhou Opera House, Guangzhou, China, 2010

伊夫琳·格雷斯学院，伦敦，英国，2010
Evelyn Grace Academy, London, UK, 2010

银河Soho，北京，中国，2012
Galaxy Soho, Beijing, China, 2012

伦敦水上运动中心，伦敦，英国，2011
London Aquatics Center, London, UK, 2011

蛇形画廊，伦敦，英国，2013
Serpentine Sackler Gallery, London, UK, 2013

牛津大学圣安东尼学院中东文化中心，英国，2015
Oxford University Middle East Center at St Antony's College, UK, 2015

梅斯纳尔山皇冠博物馆，意大利，2015
Messner Mountain Museum Corones, Italy, 2015

东大门设计广场（DDP），首尔，韩国，2013
Dongdaemun Design Plaza (DDP), Seoul, South Korea, 2013

Heydar Aliyev文化中心，巴库，阿塞拜疆，2012 / Heydar Aliyev Center, Baku, Azerbaijan, 2012

Sberbank科技园,莫斯科,俄罗斯,2016—2019
Sberbank Technopark, Moscow, Russia, 2016~2019

马萨里克火车站,布拉格,捷克共和国,2016—2022 / Masaryk Railway Station, Prague, Czech Republic, 2016~2022

这一总面积为8500m²的艺术中心是在2003年对公众开放的。其设计理念便是打造一个动态的公共空间。它被誉为"城市的地毯画,从建筑的外观开始,穿过入口和大堂,再到室内空间,在周边区域形成人流区"。[15] 扎哈将该中心的空间体验描述为"一种多角度视角和分散型几何空间所构成的新型空间流动性,以此来体现现代生活的混沌的流动性。"[15]

在上述所提到的三个项目——香港山顶休闲俱乐部、德国莱茵河畔维特拉消防站以及俄亥俄州辛辛那提的理查德&洛伊丝·罗森萨尔当代艺术中心中,扎哈追求空间的流动设计,以使建筑能够和城市的背景以及景观结合起来。在这些早期的项目中,扎哈的兴趣在于将建筑、景观和地形结合在一起。建筑形式和背景之间的关系是她之后职业生涯作品中最主要的特点。

在完成俄亥俄州辛辛那提的理查德&洛伊丝·罗森萨尔当代艺术中心后大约十五年的时间里,扎哈·哈迪德完成了好几个雄心勃勃的突破性项目。在2001年的法国斯特拉斯堡的Hoenheim-Nord总站及停车场项目中,她挑战了绘图和建筑的界限,或者是挑战了二维和三维之间的界限。画在停车场地面上的线条无缝转接成垂直性的元素(如柱子)或其他水平性元素(天篷)。在这一项目中,区分体量的线条和阴影的扁平平面还是有难度的。因斯布鲁克的滑雪跳台(奥地利,2002)是扎哈设计的一个与众不同的项目。这个90m长、50m高的跳台的特点是半塔半桥的结构,该结构以宏伟的阿尔卑斯山的天际线为背景。同样的,为Zaragoza(西班牙)的国际展览会而设计的桥形展馆,其特点是在埃布罗河两岸搭建了一个跨度为155m的桥梁结构来容纳展览场所。展馆的微微弯曲的外形正好响应了世博会的主题"水源与可持续发展"。建筑师尤为注意项目的表皮设计,覆层材料的选取受到鲨鱼鳞片的启发。正如建筑师所解释的那样,覆层的"嵌板叠加在一起,采取的是一种有机的、编织的形式,形成透视的布局,营造了一种自然的室内微型气候"。[17]

Rosenthal Center for Contemporary Art Center in Ohio was finished, Zaha Hadid realised several ambitious and groundbreaking projects. The Hoenheim-Nord Terminus and Car Park (Strasbourg, France, 2001) challenges the boundaries between drawings and architecture, or, between two and three dimensions. The lines painted on the ground of the carpark seamlessly turn into vertical elements (e.g. columns) or other horizontal parts (the canopy). In this project, it becomes difficult to differentiate lines from volumes and flat surfaces from shadows. The Bergisel Ski Jump in Innsbruck (Austria, 2002) is a clear example of the distinctive buildings designed by Hadid. This 90m long and 50m high sky jump facility is characterised by a part-tower and part-bridge structure that emerges from the suggestive Alpine skyline in the background. Similarly, the Bridge Pavilion designed for the International Exhibition held in Zaragoza (Spain) in 2008 merges two building typologies, featuring a spectacular exhibition space within a bridge structure spanning 155 meters between the two banks of the Ebro River. The gently curving outline of the pavilion relates to the Expo's theme "Water and Sustainable Development". Special attention has been given to the skin of the project, materialised with a cladding inspired by shark scales. As explained by the architects, the cladding "*is laid out in an optical pattern of superimposed panels, an organic, braided form that creates a natural microclimate on the interior.*"[17]

Parametricism

Whilst the early projects were characterised by a continuous challenge to the pre-set rules and conventions of architecture, the more recent work of Zaha Hadid embraced the use of sophisticated digital tools. Parametric design has been increasingly important in Zaha Hadid's work. On the one hand, the complex geometries designed within the office required a growing degree of organisation and control. On the other hand, the influence of people like Patrick Schumacher (who joined the practice in 1988) has been increasingly significant in Zaha Hadid's work. Schumacher's theorisation of parametricism (a term that he coined in 2008) has been published widely and promoted as "*a new global style for architecture and urban design*".[18]

As Zaha Hadid Architects practice has been innovative in working with space, shapes and forms in the 1980s and 1990s, the office represents today a global reference for designers interested in digital and parametric design. The ZHA | CODE – a group of coders, programmers and experts in software design, algorithm writing and scripting working within Zaha

参数化主义

扎哈早期项目的主要特点是不断地挑战建筑行业已形成的规则和惯例,而她近期的更多作品则是使用复杂的电子工具。参数设计在扎哈的工作中越来越重要。一方面,随着办公室设计中更具组织力和控制力要求的出现,设计师们需要构造出更复杂的几何形状。另一方面,像帕特里克·舒马赫(在1988年加入事务所)这样具有影响力的人,在扎哈作品中起的作用也越来越重要。舒马赫的参数设计方面的理论(他在2008年发明的新词语)被广泛地应用,并且被推广为"全球建筑和城市设计的新潮流"。[18]

在20世纪80年代和90年代,随着扎哈事务所在空间、形状和形式方面越来越具有创新力,他们为全球的数字和参数设计爱好者树立了风向标。ZHA|CODE事务所由一群在软件设计方面的编码者、程序员以及专家组成,与扎哈事务所的主要人员一起编写程序和脚本,是为大部分扎哈近期作品试验其可行性的核心团队。CODE计算机设计研究组研发新工具和新技术,用于找形以及形状、构件和系统的合理规划,这些如今都是扎哈作品的闪光点。此外,该团队还参与和参数化设计有关的辩论、全球性会议(如建筑辅助设计计算机协会)和展览(例如威尼斯双年展[19]),做出了很大的贡献。

先进的参数化设计已用在哈迪德最近的几个项目中,包括香奈儿的现代艺术展馆,通过构建一个面积为700m²的移动结构来为2010年法国时装公司的开幕展会提供场所,该项目通过一系列连续的拱形构件创建了一个蜿蜒且光滑的曲形几何体,并且覆以反光材料板。根据每个展览的要求,内部空间被赋予了最大程度的灵活性。

2010年,扎哈·哈迪德的三个重点项目举行了落成仪式。意大利罗马的MAXXI——21世纪艺术博物馆是经过极大的争议和拖延后才建造完成的。这一项目的参赛历史可以追溯到1998年,体现了许多扎哈·哈迪德早期项目的设计特点。正如设计师所描述的那样,"MAXXI颠覆了以往博物馆的概念,它不再只是一个固定的建筑实体,相反,它是一

Hadid's main office – is the core team responsible for the feasibility of most of the latest Hadid projects. The computational design research group CODE develops new tools and technologies for form-finding and rationalisation of the shapes, elements, and systems that characterise Hadid's work today. Moreover, the group contributes largely to the ongoing debates on parametric design, participating in global conferences (e.g. ACADIA – the Association for Computer Aided Design in Architecture), and exhibitions (e.g. Venice Biennale[19]).

Highly-advanced parametric design has been used in several of the recent Hadid projects, including the Chanel Contemporary Art Pavilion, a 700m² mobile structure that houses an exhibition space inaugurated in 2010 for the French fashion company. The project displays a series of continuous arch-shaped elements creating a sinuous and smooth curvilinear geometry clad with reflective material panels. The space inside allows for a maximisation of the spatial flexibility according to each exhibition's requirements.

In 2010, three of Zaha Hadid's key projects were inaugurated. The MAXXI – Museum of 21st Century Arts in Rome, Italy, was finished after great controversy and delays. The project, whose competition entry dates back to 1998, retrieves many of the design ideas characteristic of Zaha Hadid's early projects. As described by the architects the "*MAXXI supersedes the notion of museum as object or fixed entity, presenting instead a field of buildings accessible to all, with no firm boundary between what is within and what without. Central to this new reality [...] is a confluence of lines – walls that constantly intersect and separate to create indoor and outdoor spaces.*"[20] The concept of fluidity and spatial experience characterises the multiple perspective points of this innovative museum space.

The imposing Guangzhou Opera House in China and the Evelyn Grace Academy in Brixton, London were also finished in 2010. The language of fluidity and seamlessness distinctive of the architecture of Zaha Hadid are clear in both projects. In the former, the new cultural facility creates a dialogue between architecture, dense built environment and the surrounding landscape. The architects explain that "*the design evolved from the concept of a natural landscape and the fascinating interplay between architecture and nature, engaging the principles of erosion, geology and topography.*"[21] In the latter, the Z-shaped new academy building is located on the diagonal of the site, giving a very strong urban character to the new facilities in this populated residential London neighbourhood. One can find similarities between the position of main volumes

个可以进出的领域,不再被局限于内部应该有什么或者不该有什么。这一现实性设计的核心……是线条设计的交汇。通过墙壁之间的结合和分离,建筑师创造出室内和室外的空间"。[20] 流动性和空间体验使这一创新型博物馆具有了多重视觉空间。

宏伟的中国广州歌剧院、伦敦的伊夫林格雷斯学院也是在2010年竣工的。扎哈对于建筑的流动性设计语言以及无缝连接的设计特色在这两个项目中都有所体现。就流动性而言,新的文化设施在建筑、密集的建筑环境以及周边景观之间建立起了一种对话。建筑师解释到:"设计主要来源于自然景观的概念和建筑与自然之间的有趣互动,同时考虑到侵蚀情况、地质以及地貌。"[21] 就无缝连接的特色而言,Z形的新学术楼坐落在场地的对角线位置,赋予位于伦敦人口稠密区的新设施一种很强的都市感。我们可以发现在主体量位置的选取上,伊夫林·格雷斯学院与MAXXI博物馆在其与环境的关系方面有着共同之处,它们相对于直角的城市肌理来说都是倾斜的。

扎哈·哈德迪在伦敦建造的颇受赞扬的项目之一是2012年伦敦奥运会的水上活动中心。这座建筑主要的特点是采用了弯曲的几何构造。其灵感主要来源于水的运动。起伏的双曲面屋顶被设计得如同横扫地面的波浪一般,将中心的泳池围合起来。泳池和大型领奖台(内设所有功能性构件)之间是一个光滑的玻璃曲面,阳光可以通过玻璃曲面照射进来,玻璃曲面在视觉上创造了一种室内与室外空间上的连续性。

同样地,在2012年阿塞拜疆巴库竣工的Heydar Aliyev文化中心也有着白色的波状外表,并且周围广场和主建筑的室内之间还建立了一种连续的流动关系。广场原来是位于沉闷的苏联城市肌理中的一个体量,如今它非常宏伟,将公共室内空间包裹其中。通过起伏、分叉、褶皱、弯曲等设计处理,广场成为建筑景观。正如建筑师所解释的那样,"这一造型使建筑物和城市景观、建筑外观和城市广场、建筑和背景、内部和外部之间的传统分化不再是那么明显。"[22]

1. http://www.pritzkerprize.com/2004/bio
2. http://worldarchitecture.org/articles-links/cefmz/architects-and-designers-pay-tributes-to-zaha-hadid.html/ accessed: 20 May 2016
3. http://www.fosterandpartners.com/news/archive/2016/04/dame-zaha-hadid/ accessed: 15 May 2016
4. https://www.theguardian.com/artanddesign/2016/mar/31/star-architect-zaha-hadid-dies-aged-65/ accessed: 10 May 2016
5. Twitter of Studio Libeskind @DanielLibeskind 31 Mar 2016
6. http://edition.cnn.com/2016/03/31/europe/architect-zaha-hadid-dead/ accessed: 10 May 2016
7. http://aasarchitecture.com/2016/04/zaha-hadid-1950-2016-great-architects-most-iconic-buildings.html/ accessed: 10 May 2016
8. http://www.designboom.com/architecture/lord-norman-foster-tribute-zaha-hadid-statement-04-01-2016/ accessed: 10 May 2016
9. http://www.dezeen.com/2016/04/01/kengo-kuma-comments-tribute-zaha-hadid-tokyo-2020-stadium/ accessed: 11 May 2016
10. http://www.dezeen.com/2016/04/01/rem-koolhaas-exclusive-interview-friendship-zaha-hadid-beauty-strength/ accessed: 11 May 2016
11. http://www.zaha-hadid.com/ accessed: 20 May 2016
12. Quotation from: Michael Mönninger, Fire Station, Weil am Rhein, Domus #752, September 1993
13. http://www.zaha-hadid.com/architecture/the-peak-leisure-club/ accessed: 11 May 2016

of the Evelyn Grace Academy and the MAXXI with regard to their context, where the main buildings are oblique to the orthogonal urban fabric.

One of the most acclaimed projects that Zaha Hadid built in London is the Aquatics Center for the London Olympics in 2012. The building is characterised by a sinuous geometry deeply inspired by the idea of the water in motion. An undulating double-curvature roof sweeps up from the ground as a wave, enclosing the pools at the center. Between this and the large podium that accommodates all the programmatic elements, an imposing glazed curved facade brings natural light inside, while creating a continuous visual relationship between interior and exterior.

Similarly, the wavy and gentle white surface of the Heydar Aliyev Cultural Center, finished in 2012 in Baku, Azerbaijan, establishes a continuous and fluid relationship between its surrounding plaza and the interior of the main building. The plaza – formerly a void on the heavily structured Soviet urban fabric – rises majestically to embrace the public interior space. Through undulations, bifurcations, folds and inflections the plaza becomes the architectural landscape. As explained by the architects *"with this gesture the buildings blur the conventional differentiation between architectural object and urban landscape, building envelope and urban plaza, figure and ground, interior and exterior."*[22]

The Dongdaemun Design Plaza(DDP) in Seoul, South Korea was designed as cultural hub working 24/7 in the historic and vibrant district of the South Korean capital. Completed in 2014, the 86,000m² complex generates a series of hybrid yet continuous spaces that combine the busy urban activities of the district and the hilly backdrop of Seoul with the captivating interior, characterised by continuous changes in height and length of the visual perspectives. The project *"integrates the park and plaza seamlessly as one, blurring the boundary*

韩国首尔东大门设计广场 (DDP) 是一个文化交汇的地方，位于韩国首都的一处古老但充满活力的区域，可全天候为公众服务。这座广场于2014年竣工，面积为86 000m²，为城市带来了一系列复杂但连续的空间，这些空间将该区域繁忙的城市活动和首尔的丘陵背景与吸引人眼球的室内空间结合起来，在视觉上，其高度和长度不断发生着变化。这一项目"使公园和广场之间实现了无缝连接，融为一体，打破了建筑和自然在空间上的界限，形成一个连续的、流动的景观……面积为30 000m²的公园对韩国传统的花园设计进行了全新的解释：进行分层以及水平设计，无需一个主导人们视野的独立特点，内部和外部空间的界限变得模糊。"23

转折点
用舒马赫的话说24，扎哈使建筑领域得到了"前所未有的拓展"。我们认为这一点体现在许多层面。首先，她通过自己的项目向我们展示了如何在实际施工中挑战思维的界限，转移建筑项目中再现、施工以及操作之间的界限。其次，她在我们对建筑空间的理解方面也起到了巨大的作用。在她的项目中，她增加了现代建筑中的空间变化形式，拓展了建筑师在规划设计中可能的排列方式。再次，她的建筑增加了我们对于空间、体量和表面的认知。她所创造的空间性很好地佐证了新空间设计存在无限的可能性。在她的建筑中，角度、线条、边缘都被流畅无缝的表面以及连续的曲线所取代。最后，扎哈成为天赋、激情以及自我激励这些特点的强有力的代言人，年轻的新生代建筑师都将（可能将）扎哈作为自己在建筑事业上的追求目标。扎哈·哈德迪将代表建筑史上的一个转折点，她之后的建筑、空间设计与她生前的相比，发生了变化。

---

14. https://www.moma.org/momaorg/shared/pdfs/docs/press_archives/6526/releases/MOMA_1988_0029_29.pdf?2010/ accessed: 20 May 2016
15. http://www.zaha-hadid.com/architecture/lois-richard-rosenthal-center-for-contemporary-art/ accessed: 11 May 2016
16. http://blog.unitee.eu/meet-the-new-europeans/on-urban-design-architecture-and-pushing-boundaries-meet-zaha-hadid-our-new-european-of-the-month/ accessed: 18 May 2016
17. http://www.zaha-hadid.com/design/zaragoza-bridge-pavilion/ accessed: 20 May 2016
18. Parametricism - A New Global Style for Architecture and Urban Design. Published in: AD Architectural Design - Digital Cities, Vol 79, No 4, July/August 2009
19. Cf. http://www.zha-code-education.org/ ZHA CODE & Collaborators feature in Zaha Hadid Exhibition. Venice Biennale 2016./ accessed: 20 May 2016
20. http://www.zaha-hadid.com/architecture/maxxi/ accessed: 11 May 2016
21. http://www.zaha-hadid.com/architecture/guangzhou-opera-house/ accessed: 11 May 2016
22. http://www.zaha-hadid.com/architecture/heydar-aliyev-centre/ accessed: 11 May 2016
23. http://www.zaha-hadid.com/2014/03/24/dongdaemun-design-plaza-opens-21-march/ accessed: 11 May 2016
24. http://www.dezeen.com/2016/05/03/zaha-hadid-unprecedented-expansion-architectural-style-space-making-patrik-schumacher/ accessed: 20 May 2016

---

*between architecture and nature in a continuous, fluid landscape […] the 30,000m² park reinterprets the spatial concepts of traditional Korean garden design: layering, horizontality, blurring the relationship between the interior and the exterior with no single feature dominating the perspective."*23

Watershed
In Schumacher's words24 Zaha Hadid delivered an *"unprecedented expansion"* in architecture. We argue that this has happened at many levels. Firstly, with her projects she showed how to challenge the boundaries of thinking in practice, shifting the boundaries of representation, construction, and management in the project of architecture. Secondly, she had a significant impact on the understanding we all have of architectural space today. With her buildings, she increased the spatial vocabulary of contemporary buildings, expanding the array of possibilities that architects have in their design toolbox. Thirdly, her buildings augmented our perception of space, volumes and surfaces. The spatialities she created are a firm testimony to the possibility of a new space, where angles, lines and edges have been superseded by smoothness, seamless surfaces and continuous curves. Lastly, Zaha Hadid represents strong evidence of what talent, passion and self-motivation can achieve. She is (and will be) inspirational for many generation of young architects and designers who see in her exemplary life a goal to pursue. Zaha Hadid will represent a watershed in history of architecture, with buildings, spaces and architecture characterised by a before and after her. Silvio Carta, Marta González

萨勒诺海运码头
Zaha Hadid Architects

扎哈·哈德迪建筑师事务所设计的萨勒诺海运码头于2016年4月25日开幕,成为该城市的城市规划的一部分。这一项目由前任市长文森佐·德·卢卡(现在已经是坎帕尼亚地区的区长)启动,一直延续到了现在的市长文森佐·那不勒斯就任期间,这个从1993年就开始的萨勒诺计划,主要是指一系列针对城市的社会、经济与环境再生而实施的实质项目和计划。码头作为1993年计划的一部分,扎哈·哈德迪建筑师事务所在2000年的设计大赛中胜出,来设计这座新码头。

该项目坐落在公共码头,该码头延伸至萨勒诺的商业港口和船坞区。新海运码头延续了城市和海洋之间的关系,并且建立了新的联系,将萨勒诺丰富的航海传统和其历史悠久的城市肌理以及构筑城市边界的远处山丘相结合。

码头就像牡蛎一样,其坚硬的、不对称的外壳保护着它内部的柔软的元素,也保护乘客在旅游旺季免遭地中海烈日的侵袭。新海运码头由三个主要的相互连接的部分组成:针对边境管制和航线而设立的行政办公室、针对国际渡轮和游轮而设立的码头以及针对地方和区域渡轮而设立的码头。

随着来自城市的游客靠近码头,码头周围的区域向上升起,呈现出一条通往室内的斜坡,斜坡将游客带到上层的大型船舶和渡轮乘船层。码头的室内布局指引游客通往一系列的内部空间。这些空间之间具有流动性,都围绕一些中心点而设,如餐厅和候车室。

在地方和区域间往来的渡轮游客可以很快地通过码头,到达水平地面,然后通过斜坡向上行至船的入口。国际渡轮和游轮的乘客可以一气呵成地完成登机手续办理、护照检查、安检和海关管制等一系列手续。抵达的乘客也须遵循相似的进程,进入码头以及行李托运区。

在夜晚,海港入口附近的码头入口发出的光芒如同灯塔一般,欢迎来到这座城市的游客。

当新的码头运行以后,无论是从功能上还是视觉上,都让人觉得它是陆地和海洋之间的平稳过渡,是一个在固态元素和液态元素之间建造的滨水结构。

人们站在码头的露台和窗户可看见壮阔的阿玛尔菲海岸、萨勒诺湾和齐伦托地区的美景。波西塔诺、卡普里、帕埃斯图姆和庞贝也在附近。这个码头将大大提高游客到该区域著名文化景点、海岸线和乡村

的可达性和旅游体验。

　　新的萨勒诺海运码头将使萨勒诺港的渡船和游轮停靠的数量大大增加，平均每年可多增加50万乘客，为该城市的服务业、零售业创造了两千多个新的就业岗位。

Salerno Maritime Terminal

Inaugurated on 25 April 2016, the new Salerno Maritime Terminal by Zaha Hadid Architects is integral to the city's urban plan. Begun by Mayor Vincenzo De Luca, now Governor of the Campania Region, and continued under the city's current Mayor Vincenzo Napoli, the 1993 plan for Salerno targeted the development of essential projects and programs for the social, economic and environmental regeneration of the city. As part of the 1993 plan, Zaha Hadid Architects won the international competition in 2000 to design the new terminal. Located on the public quay that extends into Salerno's working harbor and marina, the new maritime terminal continues the city's relationship with the sea and establishes new links; connecting Salerno's rich maritime traditions with its historic urban fabric and beyond to the hills that frame the city. Like an oyster, the terminal's hard, asymmetric shell protects the softer elements within, sheltering passengers from the intense Mediterranean sun during the popular tourist season. The new maritime terminal is composed of three primary interlocking components: administration offices for national border controls and shipping lines, the terminal for international ferries and cruise ships from around the world, and the terminal for the local and regional ferries.

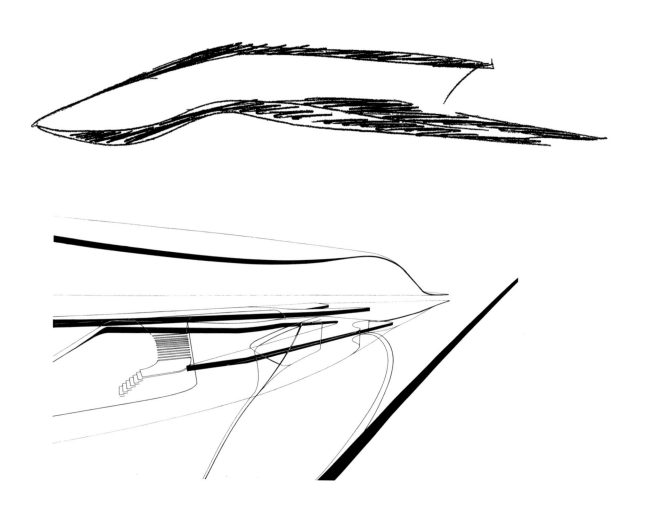

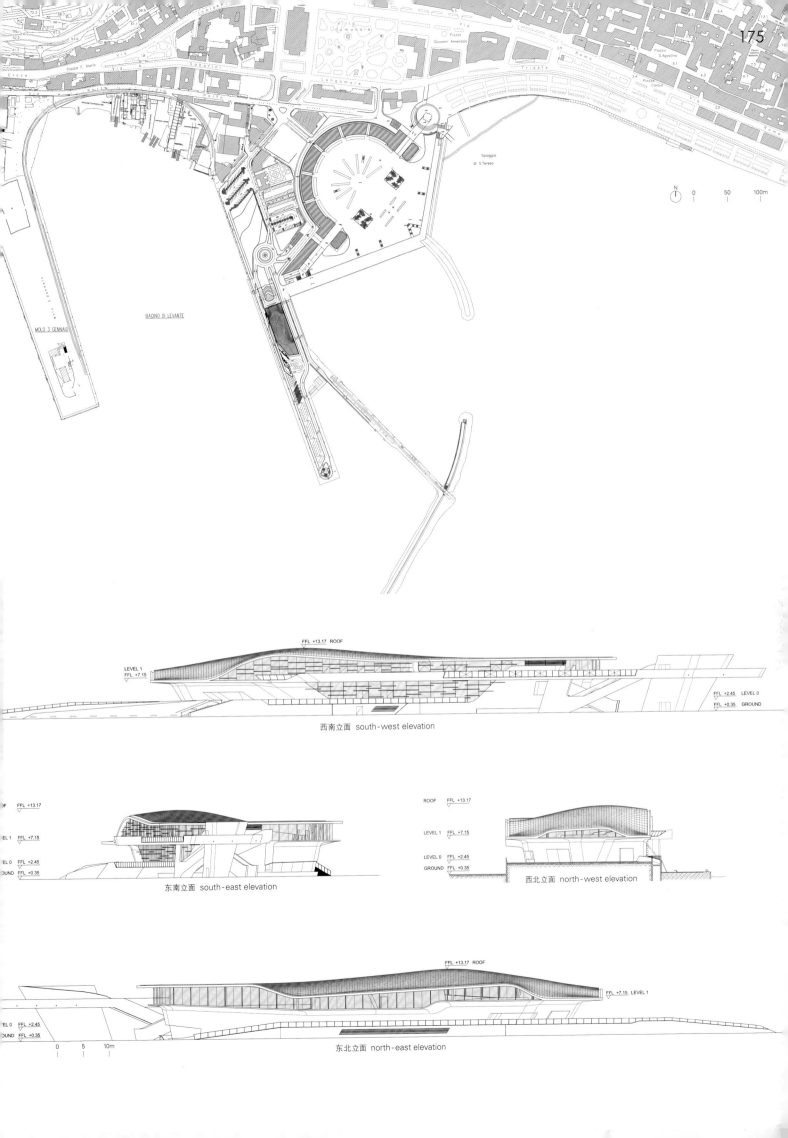

西南立面 south-west elevation

东南立面 south-east elevation

西北立面 north-west elevation

东北立面 north-east elevation

1. access ramp	6. conveyor belt–arrivals	11. informations	16. mooring quay	21. boarding bridge
2. entrance	7. arrivals lounge	12. bar	17. plant rooms	22. offices
3. conveyor belt–check in	8. custom	13. w.c.	18. arrivals hall	23. surgery
4. check in	9. arrivals ramp	14. storage	19. departures hall	24. control room
5. departures ramp	10. ticketing	15. shaft	20. passport control	25. luggage reclaim

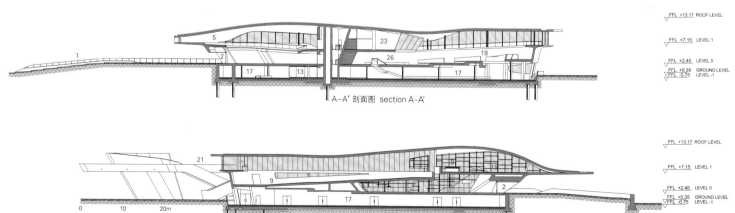

A-A' 剖面图 section A-A'

B-B' 剖面图 section B-B'

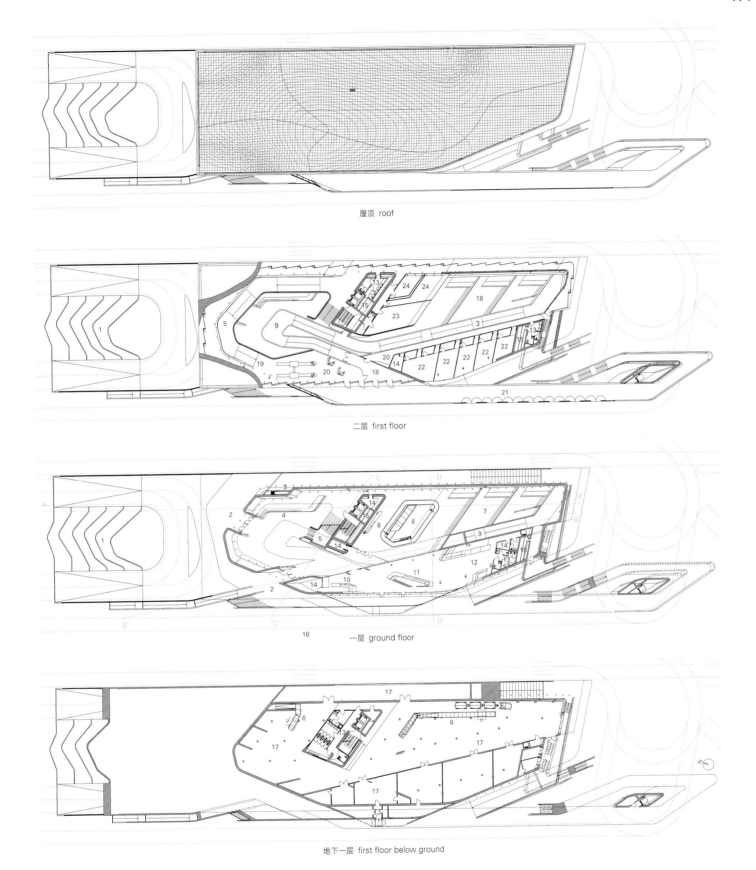

屋顶 roof

二层 first floor

一层 ground floor

地下一层 first floor below ground

C-C' 剖面图 section C-C'

D-D' 剖面图 section D-D'

The quayside gently rises as passengers approach the terminal from the city, indicating the gradually sloping path of ramps within the building which raise passengers to the embarking level of large ships and ferries. The terminal's interior arrangement orientates and leads passengers through a sequence of interior spaces that flow into each other and are organized around focal points such as the restaurant and the waiting lounge.

Local and regional ferry commuters move through the terminal quickly, arriving on the ground level and ascending via ramps to reach the upper vessel entrance. Passengers travelling on international ferries and cruise ships are guided seamlessly through check-in, passport, security and customs controls to their ship. Arriving passengers follow a similar progression through the terminal with the inclusion of the luggage reclaim area.

At night, the glow of the terminal near the harbour entrance will act as a lighthouse, welcoming visitors to the city.

The new terminal operates, both functionally and visually, as a smooth transition between land and sea, a coastal land formation that mediates between solid and liquid.

From its terraces and windows, the terminal offers spectacular views of the Amalfi Coast, the Gulf of Salerno and the Cilento. Positano, Capri, Paestum and Pompei are also nearby. The new terminal will greatly improve the accessibility and experience for visitors to the region's renowned cultural attractions, coastline and countryside.

The new Salerno Maritime Terminal will enable the port of Salerno to increase arrivals of ferry and cruise ships by 500,000 additional passengers each year, which would create up to 2,000 new jobs in the city's services and retail sectors.

功能：maritime terminal for local, regional and international ferries and cruise liners from around the world. The terminal includes waiting lounge; check-in; security control, passport control; luggage reclaim areas; quarrentine, security and administration offices; retail spaces.
用地面积：4,600m² / 有效楼层面积：4,500m² / 新建室外面积：1,600m²
总建筑规模：3 stories (one story below ground)
最高高度：13.5m / 最长长度：97m / 室内斜坡悬挑长度：20m / 琢面混凝土的总表面面积：12,000m²
玻璃的总表面面积：2,260m² / 现场浇注混凝土总体量：9,200m³
现场浇注的混凝土屋顶体量：936m³
竣工时间：2016.4

项目名称：Salerno Maritime Terminal / 地点：Salerno, Italy
建筑师：Zaha Hadid Architects / 项目建筑师：Paola Cattarin
设计团队：Vincenzo Barilari, Andrea Parenti, Anja Simons, Giovanna Sylos Labini, Cedric Libert, Filippo Innocenti, Paolo Zilli, Lorenzo Grifantini, Electra Mikelides, Eric Tong
竞赛团队：Paola Cattarin, Sonia Villaseca, Christos Passas, Chris Dopheide
当地执行建筑师：Alessandro Gubitosi_Interplan Seconda / 估算师：Pasquale Miele_Building Consulting
结构工程师：Francesco Sylos Labini_Ingeco, Sophie Le Bourva_Ove Arup & Partners (prelim. design)
M&E工程师：Roberto Macchiaroli_Macchiaroli and Partners, Felice Marotta_Itaca srl, Ove Arup & Partners (prelim. design)
海事/运输工程师：Greg Heigh_Ove Arup & Partners (London) / 照明设计师：Mark Hensmann_Equation Lighting Design (London)
客户：Comune di Salerno
摄影师：©Hélène Binet (courtesy of the architect) (except as noted)

>>46

Mirko Franzoso
Was born in 1978 in Cles, in Trento, Italy. Graduated from the School of Architecture of Venice in 2005. After collaborating with several architecture and design studies, he began an independent career in 2009 dealing with architectural design at different scales, from the preliminary phase to the executive. The fields of interest, ranging from restructuring to architectural design, from design to restoration. Over the years he took part successfully in several international architecture and design competitions.

>>22

Foster + Partners
Is an international studio for architecture, engineering and design, led by Founder and Chairman Norman Foster and a Partnership Board. Founded in 1967, the practice is characterized by its integrated approach to design, bringing together the depth of resources required to take on some of the most complex projects in the world. Over the past five decades the practice has pioneered a sustainable approach to architecture and ecology through a strikingly wide range of work, from urban masterplans, public infrastructure, airports, civic and cultural buildings, offices and workplaces to private houses and product design. The studio has established an international reputation with buildings such as the world's largest airport terminal at Beijing, Swiss Re's London Headquarters, Hearst Headquarters in New York, Millau Viaduct in France, the German Parliament in the Reichstag, Berlin, The Great Court at London's British Museum, Headquarters' for HSBC in Hong Kong and London, and Commerzbank Headquarters in Frankfurt.

>>126

Markus Schietsch Architekten
After graduation from ETH Zürich in 2002, Markus Schietsch worked in different architectural firms in New York, Vienna and Zürich before founding his own office Markus Schietsch Architekten GmbH in 2005 in Zürich. Beside the conception and realisation of various buildings Markus Schietsch is frequently giving lectures, teaches as a visiting professor and is often invited as a critic at various universities. Markus Schietsch lives and works in Zürich.

>>96
Reiulf Ramstad Architects
Reiulf Ramstad was born in 1962 in Oslo, Norway. Studied at the University of Genoa and received Ph.D. from the University of Venice(IUAV). Founded his own office in 1995 and earned a reputation for creating bold, simple architecture with a strong connection to the Scandinavian context and impressive landscape in particular. Was nominated for the Mies van der Rohe Award in 2007, 2011, 2012, 2013, 2014 and won the WAN award for leading architects of the 21st Century in 2012. His main activities related to education and research at Norwegian School of Science and Technology.

>>142
Jensen & Skodvin Architects
Jan Olav Jensen[top left] and Børre Skodvin[top right] graduated from the Oslo School of Architecture and Design and established Jensen & Skodvin Architects in 1995. They completed a variety of projects, from furniture to urban planning for public as well as private clients during these 18 years. Since 2013 Torunn Golberg[lower left] and Torstein Koch[lower right] have become partners in the office. Jan Olav Jensen received the Aga Khan Award for Architecture in 1998 and the Swedish Prince Eugen Medal in 2006. Børre Skodvin was the master of the steel workshop at the Oslo School of Architecture and design and the vice president of Norwegian Architects association.

University of Stuttgart
Is an internationally leading institution in research that integrates architecture, engineering and natural sciences. Its long tradition, most prominently represented by the work of last year's Pritzker awardee Frei Otto, and its continuing innovation in this field has recently led to establishing the Collaborative Research Center "Biological Design and Integrative Structures", funded by the German Research Council. The Research Pavilion is developed and produced by ICD(Institute for Computational Design – Prof. Achim Menges) and ITKE(Institute of Building Structures and Structural Design – Prof. Jan Knippers). This year's pavilion is a demonstrator building of the sub-project "Segmented Shell Structures".

>>78

MX_SI
Was founded by Boris Bezan, Hector Mendoza and Mara Partida in Barcelona, 2005. Focuses its work primarily on developing public architectural competitions of a cultural and historic nature. Moves on the basis of the continuous search for the essence of architecture based on the dialogue between structure, geometry, spatial and visual context and expression. Also shows great sensitivity with regards time and the place, integrating the new with the existing while highlighting the visual qualities of their projects.

>>114

Architecture-Studio
Was founded in Paris in 1973. Works with large international groups and develops big projects including large residential development. Has intensified its activities on the international stage. And the international presence is particularly strong in China. They have two permanent subsidiaries in Shanghai and Beijing. Define architecture as "an art committed with society, the construction of the surroundings of mankind". Its foundations lie on work group and shared knowledge, with the will to go beyond individuality for the benefit of dialogue and confrontation. Thus, the addition of individual knowledge turns into wide creative potential.

©Gaston Bergeret

Susanne Keneddy
Has written about design and architecture around the world for about 16 years for many international publications. Lives in Australia when not traveling.

Silvio Carta
Ph.D. (2010, University of Cagliari, Italy), Doctor Europaeus, architect and researcher based in London. His main fields of interest are architectural design and design theory. His studies have focused on contemporary architecture, digital design, architectural criticism, research through making, and analysis of the design process. Has taught at the University of Cagliari (Italy), Willem de Kooning Academy (University of Rotterdam) and Delft University of Technology, Department of Public Building. Is now senior lecturer at the University of Hertfordshire, where is investigating the potentialities of digital design applied to public space.

Marta González
Is an architect based in London with over 12 years' experience working on a number of projects including commercial, mixed use and residential schemes in Europe and Asia. Her broad knowledge and expertise covers conceptual design, detail design and construction documentation. Is currently working on a high rise residential and five star hotel development in central London. Prior moving to the UK she worked in The Netherlands where she combined her research and teaching activities with the architecture practise. Her articles have been published in several international architecture magazines.

>>62
Tectoniques Architects
With 20 years of professional experience, Tectoniques is collective in structure, comprising two generations of architects and engineers. New young associates in 2007 and 2011 brought about a movement of renewal, in an ongoing synergy of experience and dynamism. Specializes in dry construction techniques. Tectoniques has for long been interested in the principles associated with timber frame buildings. It currently has some 20 members who form organisational structures, centred on creative processes that are collegial and interactive. Lucas Jollivet and Raphael Verboud from the left.

>>36
Christiansen Andersen
Jesper Kort Andersen[right] and Mikkel Kjærgrd Christiansen[left] have a small collaborative practice based in Copenhagen. Mikkel Kjærgrd Christiansen was born in 1984 and received Master of Architecture from Royal Academy of Arts in 2011. Has worked at the Utzon Architects, Entasis Arkitekter and Lundgaard & Tranberg Arkitekter. Is currently teaching at the Royal Academy of Arts since 2010. Jesper Kort Andersen was born in 1974 and received Master of Architecture from Royal Academy of Arts in 2003. Has worked at the Klar Arkitekter, Holscher Arkitekter, Entasis Arkitekter, Polyform Arkitekter and Gottlieb Paludan Architects.

© 2017大连理工大学出版社

版权所有·侵权必究

图书在版编目(CIP)数据

木材：诗意与实用：汉英对照 / 福斯特建筑事务所等编；刘文静译. — 大连：大连理工大学出版社，2017.3
（建筑立场系列丛书）
ISBN 978-7-5685-0746-2

Ⅰ. ①木… Ⅱ. ①福… ②刘… Ⅲ. ①建筑材料－木材－汉、英 Ⅳ. ①TU531.1

中国版本图书馆CIP数据核字(2017)第056012号

出版发行：大连理工大学出版社
　　　　　（地址：大连市软件园路80号　邮编：116023）
印　　刷：上海锦良印刷厂
幅面尺寸：225mm×300mm
印　　张：11.75
出版时间：2017年3月第1版
印刷时间：2017年3月第1次印刷
出 版 人：金英伟
统　　筹：房　磊
责任编辑：许建宁
封面设计：王志峰
责任校对：高　文
书　　号：978-7-5685-0746-2
定　　价：258.00元

发　行：0411-84708842
传　真：0411-84701466
E-mail：12282980@qq.com
URL：http://www.dutp.cn

本书如有印装质量问题，请与我社发行部联系更换。

C3建筑立场系列丛书01:
墙体设计
ISBN: 978-7-5611-6353-5
定价: 150.00元

C3建筑立场系列丛书02:
新公共空间与私人住宅
ISBN: 978-7-5611-6354-2
定价: 150.00元

C3建筑立场系列丛书03:
住宅设计
ISBN: 978-7-5611-6352-8
定价: 150.00元

C3建筑立场系列丛书04:
老年住宅
ISBN: 978-7-5611-6569-0
定价: 150.00元

C3建筑立场系列丛书05:
小型建筑
ISBN: 978-7-5611-6579-9
定价: 150.00元

C3建筑立场系列丛书06:
文博建筑
ISBN: 978-7-5611-6568-3
定价: 150.00元

C3建筑立场系列丛书07:
流动的世界: 日本住宅空间设计
ISBN: 978-7-5611-6621-5
定价: 200.00元

C3建筑立场系列丛书08:
创意运动设施
ISBN: 978-7-5611-6636-9
定价: 180.00元

C3建筑立场系列丛书09:
墙体与外立面
ISBN: 978-7-5611-6641-3
定价: 180.00元

C3建筑立场系列丛书10:
空间与场所之间
ISBN: 978-7-5611-6650-5
定价: 180.00元

C3建筑立场系列丛书11:
文化与公共建筑
ISBN: 978-7-5611-6746-5
定价: 160.00元

C3建筑立场系列丛书12:
城市扩建的四种手法
ISBN: 978-7-5611-6776-2
定价: 180.00元

C3建筑立场系列丛书13:
复杂性与装饰风格的回归
ISBN: 978-7-5611-6828-8
定价: 180.00元

C3建筑立场系列丛书14:
企业形象的建筑表达
ISBN: 978-7-5611-6829-5
定价: 180.00元

C3建筑立场系列丛书15:
图书馆的变迁
ISBN: 978-7-5611-6905-6
定价: 180.00元

C3建筑立场系列丛书16:
亲地建筑
ISBN: 978-7-5611-6924-7
定价: 180.00元

C3建筑立场系列丛书17:
旧厂房的空间蜕变
ISBN: 978-7-5611-7093-9
定价: 180.00元

C3建筑立场系列丛书18:
混凝土语言
ISBN: 978-7-5611-7136-3
定价: 228.00元

C3建筑立场系列丛书19:
建筑入景
ISBN: 978-7-5611-7306-0
定价: 228.00元

C3建筑立场系列丛书20:
新医疗建筑
ISBN: 978-7-5611-7328-2
定价: 228.00元

C3建筑立场系列丛书21:
内在丰富性建筑
ISBN: 978-7-5611-7444-9
定价: 228.00元

C3建筑立场系列丛书22:
建筑谱系传承
ISBN: 978-7-5611-7461-6
定价: 228.00元

C3建筑立场系列丛书23:
伴绿而生的建筑
ISBN: 978-7-5611-7548-4
定价: 228.00元

C3建筑立场系列丛书24:
大地的皱折
ISBN: 978-7-5611-7649-8
定价: 228.00元

C3建筑立场系列丛书25:
在城市中转换
ISBN: 978-7-5611-7737-2
定价: 228.00元

C3建筑立场系列丛书26:
锚固与飞翔——挑出的住居
ISBN: 978-7-5611-7759-4
定价: 228.00元

C3建筑立场系列丛书27:
创造性加建: 我的学校, 我的城市
ISBN: 978-7-5611-7848-5
定价: 228.00元

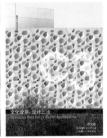
C3建筑立场系列丛书28:
文化设施: 设计三法
ISBN: 978-7-5611-7893-5
定价: 228.00元

C3建筑立场系列丛书29:
终结的建筑
ISBN: 978-7-5611-8032-7
定价: 228.00元

C3建筑立场系列丛书30:
博物馆的变迁
ISBN: 978-7-5611-8226-0
定价: 228.00元

C3建筑立场系列丛书31:
微工作 • 微空间
ISBN: 978-7-5611-8255-0
定价: 228.00元

C3建筑立场系列丛书32:
居住的流变
ISBN: 978-7-5611-8328-1
定价: 228.00元

C3建筑立场系列丛书33:
本土现代化
ISBN: 978-7-5611-8380-9
定价: 228.00元

C3建筑立场系列丛书34:
气候与环境
ISBN: 978-7-5611-8501-8
定价: 228.00元

C3建筑立场系列丛书35:
能源与绿色
ISBN: 978-7-5611-8911-5
定价: 228.00元

C3建筑立场系列丛书36:
体验与感受: 艺术画廊与剧院
ISBN: 978-7-5611-8914-6
定价: 228.00元

C3建筑立场系列丛书37:
记忆的住居
ISBN: 978-7-5611-9027-2
定价: 228.00元

C3建筑立场系列丛书38:
场地、美学和纪念性建筑
ISBN: 978-7-5611-9095-1
定价: 228.00元

C3建筑立场系列丛书39:
殡仪类建筑: 在返璞和升华之间
ISBN: 978-7-5611-9110-1
定价: 228.00元

C3建筑立场系列丛书40:
苏醒的儿童空间
ISBN: 978-7-5611-9182-8
定价: 228.00元

C3建筑立场系列丛书41:
都市与社区
ISBN: 978-7-5611-9365-5
定价: 228.00元

C3建筑立场系列丛书42:
木建筑再生
ISBN: 978-7-5611-9366-2
定价: 228.00元

| C3建筑立场系列丛书43:
休闲小筑
ISBN: 978-7-5611-9452-2
定价: 228.00元 | C3建筑立场系列丛书44:
节能与可持续性
ISBN: 978-7-5611-9542-0
定价: 228.00元 | C3建筑立场系列丛书45:
建筑的文化意象
ISBN: 978-7-5611-9576-5
定价: 228.00元 | C3建筑立场系列丛书46:
重塑建筑的地域性
ISBN: 978-7-5611-9638-0
定价: 228.00元 | C3建筑立场系列丛书47:
传统与现代
ISBN: 978-7-5611-9723-3
定价: 228.00元 | C3建筑立场系列丛书48:
博物馆: 空间体验
ISBN: 978-7-5611-9737-0
定价: 228.00元 |

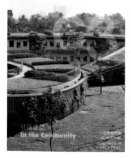     

C3建筑立场系列丛书49: 社区建筑　ISBN: 978-7-5611-9793-6　定价: 228.00元

C3建筑立场系列丛书50: 林间小筑　ISBN: 978-7-5611-9811-7　定价: 228.00元

C3建筑立场系列丛书51: 景观与建筑　ISBN: 978-7-5611-9884-1　定价: 228.00元

C3建筑立场系列丛书52: 地域文脉与大学建筑　ISBN: 978-7-5611-9885-8　定价: 228.00元

C3建筑立场系列丛书53: 办公室景观　ISBN: 978-7-5685-0134-7　定价: 228.00元

C3建筑立场系列丛书54: 城市复兴中的生活设施　ISBN: 978-7-5685-0340-2　定价: 228.00元

    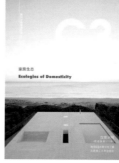 

C3建筑立场系列丛书55: 灰色建筑中的绿色自然　ISBN: 978-7-5685-0406-5　定价: 228.00元

C3建筑立场系列丛书56: 从教育角度看幼儿园建筑　ISBN: 978-7-5685-0410-2　定价: 228.00元

C3建筑立场系列丛书57: 能源意识与可持续公共空间　ISBN: 978-7-5685-0409-6　定价: 228.00元

C3建筑立场系列丛书58: 灵活的学习空间　ISBN: 978-7-5685-0439-3　定价: 228.00元

C3建筑立场系列丛书59: 家居生态　ISBN: 978-7-5685-0455-3　定价: 228.00元

C3建筑立场系列丛书60: 地方性与全球多样性　ISBN: 978-7-5685-0454-6　定价: 228.00元

C3建筑立场系列丛书61: 时间: 空间记忆　ISBN: 978-7-5685-0546-8　定价: 228.00元

C3建筑立场系列丛书62: 叩问自然之灵　ISBN: 978-7-5685-0584-0　定价: 228.00元

C3建筑立场系列丛书63: 大学建筑: 华丽的转变　ISBN: 978-7-5685-0578-9　定价: 228.00元

C3建筑立场系列丛书64: 探索瑞士建筑的异曲同工之妙　ISBN: 978-7-5685-0602-1　定价: 228.00元

C3建筑立场系列丛书65: 建筑情感: 从宗教到世俗　ISBN: 978-7-5685-0603-8　定价: 228.00元

C3建筑立场系列丛书66: 建筑情感: 创意办公　ISBN: 978-7-5685-0747-9　定价: 258.00元

出版社淘宝店

"C3建筑立场系列丛书"已由大连理工大学出版社出版，欢迎订购！

◆ 编辑部咨询电话: 许老师/0411-84708405
◆ 发行部订购电话: 王老师/0411-84708943

上架建议: 建筑设计

定价: 258.00元